Ergebnisse der Mathematik und ihrer Grenzgebiete

Band 47

Leopoldo Nachbin

Topology on Spaces of Holomorphic Mappings

Springer-Verlag New York Inc. 1969

Leopoldo Nachbin

George Eastman Professor of Mathematics
University of Rochester
Rochester, New York, USA

Pesquisador Titular
Instituto de Matemática Pura e Aplicada
Rio de Janeiro, Guanabara, Brazil

 Library of Congress Catalog Card Number 68-29710. Printed in Germany
Title No. 4591

Dedicado ao meu filho

ANDRÉ NACHBIN

Preface

The present report on spaces of holomorphic mappings was prepared for the Sexto Colóquio Brasileiro de Matemática (Poços de Caldas, Minas Gerais, Brazil, July 1967). I also had the opportunity of giving a series of lectures on this material while I was a visiting member at the Center for Theoretical Studies of the University of Miami (Coral Gables, Florida, USA, February 1968). The preparation of this report was sponsored in part by the USA National Science Foundation through a grant to the University of Rochester.

I am glad to thank Professors Paul R. Halmos and Peter J. Hilton for accepting my text as part of the series Ergebnisse der Mathematik und ihre Grenzgebiete.

Rochester, New York 1968 Leopoldo Nachbin

Contents

§ 1. Introduction

The purpose of this monograph is to describe a natural method of endowing certain vector spaces of holomorphic mappings with locally convex topologies, and to derive a few results for the sake of illustration of the simple ideas involved in such a method.

The need of the following considerations [13] was prompted by the remark that the largest natural locally convex topology to be used on $\mathscr{H}(U;F)$ (see § 2 for notation and terminology) is not the one induced on it by the compact-open topology on the vector space $\mathscr{C}(U;F)$ of all continuous F-valued functions on U (unlike what happens when E is finite dimensional and thus locally compact). A seminorm p on $\mathscr{H}(U;F)$ is said to be ported by a compact subset K of U if, given any open subset V of U containing K, we can find a real number $c(V)>0$ for which

$$p(f)\leqslant c(V)\cdot\sup_{x\in V}\|f(x)\|$$

holds for every $f\in\mathscr{H}(U;F)$. It is to be noted that f is not necessarily bounded on V; however, once $f\in\mathscr{H}(U;F)$ and the compact subset K of U are given, there is clearly an open subset V of U containing K on which f is bounded. In other words, the above estimate imposes a restriction on p for every f, although its right-hand side may occasionally be infinite. The natural topology $\mathscr{I}_\omega$ on $\mathscr{H}(U;F)$ is defined by the seminorms on $\mathscr{H}(U;F)$ that are ported by compact subsets of U. It is much larger than the topology on $\mathscr{H}(U;F)$ induced by the compact-open topology on $\mathscr{C}(U;F)$ whenever $\dim E=\infty$ and $F\neq 0$; otherwise the two topologies on $\mathscr{H}(U;F)$ coincide. The same natural method used here, namely that of estimating functions on arbitrarily small neighborhoods of fixed compact subsets (as expressed by the above estimate), can be of course applied to $\mathscr{C}(U;F)$ itself; but then it gives back the compact-open topology on $\mathscr{C}(U;F)$, and not a larger one. It is only in the case of certain vector subspaces of $\mathscr{C}(U;F)$ which are not too large, or too small, that such a method leads to a topology actually bigger than the one induced by the compact-open topology on $\mathscr{C}(U;F)$.

On the other hand, certain questions about convolution and partial differential operators, Fourier and Borel transforms, distributions, etc. in infinite dimensions (see [6] and [15] for instance) lead to important types of holomorphy, and of differentiability or real-analyticity as well, and so to corresponding spaces of mappings and their natural topologies. Important examples of such types are found in the nuclear, the integral, the Hilbert-Schmidt cases, etc., all stemming from the corresponding kinds of continuous m-homogeneous polynomials (or equivalently, except perhaps for the norms, of continuous m-linear mappings; see [5] for instance).

§ 9 introduces the concept of holomorphy type Θ from E to F; and § 11 deals with the natural topology $\mathscr{I}_{\omega,\Theta}$ on the vector subspace $\mathscr{H}_\Theta(U;F)$ of $\mathscr{H}(U;F)$. Simplifications arising in the case $\mathscr{H}(U;F)$ are described in § 14. The rest of the monograph is devoted to sampling a few results. We refrain here from being as complete in the case $\dim E=\infty$ as one would hope for in view of the existing knowledge in the case $\dim E<\infty$ (see [11] for instance, and the literature quoted there as well). Thus we do not deal at all with the standard properties of locally convex spaces to be investigated in this connection, with spaces of functions holomorphic about a fixed set not necessarily open or compact, with spaces of real-analytic or differentiable functions, or with the dislinearization of the theory through the use of Banach manifolds, etc. A thorough exposition of such aspects deserves a lengthier monograph to itself.

§ 2. Notation and Terminology

In this section, we shall summarize the main notation and terminology used in this monograph. Explicit reference is made to the pertinent section in which they are introduced.

$\mathbb{N}$, $\mathbb{R}$, and $\mathbb{C}$ will denote the systems of all natural integers, of all real numbers and of all complex numbers, respectively.

E and F will represent two complex Banach spaces.

U will denote a non-void open subset of E. We say that U is ξ-equilibrated, with respect to one of its points ξ, in case $(1-\lambda)\xi+\lambda x \in U$ for every $x \in U$ and $\lambda \in \mathbb{C}$, $|\lambda| \leqslant 1$. At any rate, the set U_ξ of all $x \in U$ such that $(1-\lambda)\xi+\lambda x \in U$ for every $\lambda \in \mathbb{C}$, $|\lambda| \leqslant 1$ is the largest open ξ-equilibrated subset of U.

The open and closed balls with center at ξ and radius ρ in a normed space will be denoted by $B_\rho(\xi)$ and $\bar{B}_\rho(\xi)$, respectively. Correspondingly, we set

$$B_\rho(X) = \bigcup_{x \in X} B_\rho(x),$$

$$\bar{B}_\rho(X) = \bigcup_{x \in X} \bar{B}_\rho(x),$$

for a subset X of a normed space.

For each $m \in \mathbb{N}$, we shall represent by $\mathscr{L}({}^mE;F)$ the Banach space of all continuous m-linear mappings from $E^m = E \times \cdots \times E$ (m times) to F; by $\mathscr{L}_s({}^mE;F)$ the closed vector subspace of all continuous symmetric m-linear mappings from E^m to F; and by $\mathscr{P}({}^mE;F)$ the Banach space of all continuous m-homogeneous polynomials from E to F. If $A \in \mathscr{L}({}^mE;F)$, we have the corresponding elements $A_s \in \mathscr{L}_s({}^mE;F)$ and $\hat{A} \in \mathscr{P}({}^mE;F)$. See § 3.

$\mathscr{C}(U;F)$ and $\mathscr{H}(U;F)$ will represent the vector spaces of all continuous and of all holomorphic mappings from U to F, respectively. If $f\in\mathscr{H}(U;F)$, we denote by

$$d^m f(x)\in\mathscr{L}_s(^mE;F),$$
$$\hat{d}^m f(x)\in\mathscr{P}(^mE;F),$$

its differential at $x\in U$, looked at as a continuous symmetric m-linear mapping and as a continuous m-homogeneous polynomial, respectively. We have correspondingly the mappings

$$d^m f\in\mathscr{H}(U;\mathscr{L}_s(^mE;F)),$$
$$\hat{d}^m f\in\mathscr{H}(U;\mathscr{P}(^mE;F)),$$

and the differentiation operators

$$d^m:\mathscr{H}(U;F)\to\mathscr{H}(U;\mathscr{L}_s(^mE;F)),$$
$$\hat{d}^m:\mathscr{H}(U;F)\to\mathscr{H}(U;\mathscr{P}(^mE;F)),$$

of order $m=0,1,\ldots$ The Taylor series of f at $\xi\in U$ is

$$f(x)\cong\sum_{m=0}^{\infty}\frac{1}{m!}d^m f(\xi)(x-\xi)^m$$
$$\cong\sum_{m=0}^{\infty}\frac{1}{m!}\hat{d}^m f(\xi)(x-\xi),$$

whereas the Taylor polynomial $\tau_{m,f,\xi}$ of order m of f at ξ is given by

$$\tau_{m,f,\xi}(x)=\sum_{\ell=0}^{m}\frac{1}{\ell!}d^\ell f(\xi)(x-\xi)^\ell$$
$$=\sum_{\ell=0}^{m}\frac{1}{\ell!}\hat{d}^\ell f(\xi)(x-\xi),$$

where $x\in E$. See § 5.

Θ will always denote a holomorphy type from E to F. See § 9.

The concepts of seminorm, topology defined by a set of seminorms, and bounded subset are meant in the sense of the theory of topological vector spaces.

§ 3. Continuous Polynomials

Definition 1. Letting $m=1,2,\ldots$, we shall denote by $\mathscr{L}(^mE;F)$ the Banach space of all continuous m-linear mappings from $E^m=E\times\ldots\times E$ (m times) to F, with respect to the pointwise vector operations and the norm defined by

$$\|A\| = \sup_{x_1\neq 0,\ldots,x_m\neq 0} \frac{\|A(x_1,\ldots,x_m)\|}{\|x_1\|\ldots\|x_m\|},$$

where $A\in\mathscr{L}(^mE;F)$ and $x_1,\ldots,x_m\in E$. Notice that

$$\|A(x_1,\ldots,x_m)\| \leqslant \|A\|\cdot\|x_1\|\ldots\|x_m\|.$$

We shall denote by $\mathscr{L}_s(^mE;F)$ the closed vector subspace of $\mathscr{L}(^mE;F)$ of all such A that are symmetric. For $A\in\mathscr{L}(^mE;F)$, we define its *symmetrization* $A_s\in\mathscr{L}_s(^mE;F)$ by

$$A_s(x_1,\ldots,x_m) = \frac{1}{m!}\sum A(x_{j_1},\ldots,x_{j_m}),$$

where summation is over the $m!$ permutations $(j_1,\ldots,j_m)$ of $(1,\ldots,m)$. For $m=0$, we shall let $\mathscr{L}(^0E;F)=\mathscr{L}_s(^0E;F)=F$ as a Banach space, and $A_s=A$ for $A\in\mathscr{L}(^0E;F)$. Notice that $\|A_s\|\leqslant\|A\|$, and that $A\mapsto A_s$ is a continuous projection of $\mathscr{L}(^mE;F)$ onto $\mathscr{L}_s(^mE;F)$, for $m=0,1,\ldots$

Definition 2. If $A\in\mathscr{L}(^mE;F)$ and $x\in E$, we shall write Ax^m to denote $A(x,\ldots,x)$, where x is repeated m times, in case $m=1,2,\ldots$; and we shall define $Ax^0=A$, in case $m=0$. A *continuous m-homo-*

geneous polynomial P from E to F is a mapping $P: E \to F$ for which there is some $A \in \mathscr{L}(^m E; F)$ such that

$$P(x) = A x^m$$

for every $x \in E$. In order to denote that P corresponds to A in this way, we shall write $P = \hat{A}$. We shall denote by $\mathscr{P}(^m E; F)$ the Banach space of all continuous m-homogeneous polynomials from E to F, with respect to the pointwise vector operations and the norm defined by

$$\|P\| = \sup_{x \neq 0} \frac{\|P(x)\|}{\|x\|^m},$$

where $P \in \mathscr{P}(^m E; F)$ and $x \in E$. Notice that

$$\|P(x)\| \leqslant \|P\| \cdot \|x\|^m$$

and that, for $m = 0$, $\mathscr{P}(^0 E; F)$ is the vector space of all constant mappings from E to F. We have $\hat{A} = \widehat{A_s}$.

Proposition 1. *The mapping*

$$A \in \mathscr{L}_s(^m E; F) \mapsto \hat{A} \in \mathscr{P}(^m E; F)$$

is a vector space isomorphism and a homeomorphism of the first onto the second Banach space. Moreover

$$\|\hat{A}\| \leqslant \|A\| \leqslant \frac{m^m}{m!} \|\hat{A}\|.$$

Proof: The proposition is trivial for $m = 0$ and $m = 1$. Let then $m \geqslant 2$. The mapping $A \mapsto \hat{A}$ is linear on $\mathscr{L}(^m E; F)$ onto $\mathscr{P}(^m E; F)$. Since $\hat{A} = \widehat{A_s}$, we see that this mapping is linear on $\mathscr{L}_s(^m E; F)$ onto $\mathscr{P}(^m E; F)$. We may verify the so-called "polarization formula"

$$A(x_1, \ldots, x_m) = \frac{1}{m! \, 2^m} \sum_{\substack{\varepsilon_1 = \pm 1 \\ \cdots\cdots \\ \varepsilon_m = \pm 1}} \varepsilon_1 \ldots \varepsilon_m \hat{A}(\varepsilon_1 x_1 + \ldots + \varepsilon_m x_m)$$

expressing $A \in \mathscr{L}_s(^m E; F)$ back from $\hat{A}$, where $x_1, \dots, x_m \in E$. This formula proves that the mapping $A \mapsto \hat{A}$ is one-to-one on $\mathscr{L}_s(^m E; F)$ and actually gives us an explicit expression for the inverse mapping $\hat{A} \mapsto A$. It is clear that $\|\hat{A}\| \leqslant \|A\|$ for $A \in \mathscr{L}(^m E; F)$. If we assume $A \in \mathscr{L}_s(^m E; F)$ and $\|x_1\| = \dots = \|x_m\| = 1$ in the polarization formula and use

$$\|\hat{A}(\varepsilon_1 x_1 + \cdots + \varepsilon_m x_m)\| \leqslant \|\hat{A}\| \cdot m^m,$$

we get

$$\|A(x_1, \dots, x_m)\| \leqslant \frac{m^m}{m!} \|\hat{A}\|,$$

hence

$$\|A\| \leqslant \frac{m^m}{m!} \|\hat{A}\|. \qquad \textit{Q. E. D.}$$

Remark 1. There are simple examples showing that

$$\frac{m^m}{m!}$$

is the best universal constant ocurring in Proposition 1. For instance, take $E = \mathbb{C}^m$ and $F = \mathbb{C}$, where E is normed by

$$\|x\| = |x_1| + \cdots + |x_m|$$

if $x = (x_1, \dots, x_m) \in E$. Define $P: E \to F$ by

$$P(x) = x_1 \dots x_m$$

if $x = (x_1, \dots, x_m) \in E$; and

$$A(x_1, \dots, x_m) = \frac{1}{m!} \sum x_{j_1, 1} \dots x_{j_m, m}$$

if $x_j = (x_{j,1}, \dots, x_{j,m}) \in E$ $(j = 1, \dots, m)$, where the summation is over all permutations $(j_1, \dots, j_m)$ of $(1, \dots, m)$ (compare with the concept of symmetrization in Definition 1). It is easily seen that

$$A \in \mathscr{L}_s(^m E; F), \qquad \|A\| = \frac{1}{m!},$$

$$P = \hat{A} \in \mathscr{P}(^m E; F), \|P\| = \frac{1}{m^m},$$

so that, actually,

$$\|A\| = \frac{m^m}{m!} \|\hat{A}\|.$$

The above example could have been arranged with a single E of infinite dimension valid for every m, namely $E = \ell^1$ (see Remark 1, § 4). If E is finite dimensional, of dimension $n < m$, the universal constant

$$\frac{m^m}{m!}$$

in Proposition 1 can be replaced by a better one depending on m and n. We refrain from going into the details.

Definition 3. A *continuous polynomial* P from E to F is a mapping $P: E \to F$ for which there are $m = 0, 1, \ldots$, $P_k \in \mathscr{P}(^k E; F)$ $(k = 0, \ldots, m)$ such that

$$P = P_0 + \cdots + P_m.$$

We shall denote by $\mathscr{P}(E; F)$ the vector space of all continuous polynomials from E to F with respect to pointwise vector operations.

Proposition 2. *If* $P \in \mathscr{P}(E; F)$, $P \neq 0$, *there is one and only one way of writing*

$$P = P_0 + \cdots + P_m,$$

with $m = 0, 1, \ldots$, $P_k \in \mathscr{P}(^k E; F)$ $(k = 0, \ldots, m)$ *and* $P_m \neq 0$.

Proof. It suffices to prove that

$$P_0 + \cdots + P_m = 0$$

imply

$$P_0 = 0, \ldots, P_m = 0.$$

In fact, if we write

$$\sum_{k=0}^{m} P_k(x) = 0$$

for every $x \in E$, replace x by λx, $\lambda \in \mathbb{C}$, divide out by λ^m if $\lambda \neq 0$ and let $\lambda \to \infty$, we get $P_m = 0$. We then get inductively $P_{m-1} = 0, \ldots, P_0 = 0$. *Q. E. D.*

§ 4. Convergent Power Series

Definition 1. *A power series from E to F about* $\xi \in E$ is a series in $x \in E$ of the form

$$\sum_{m=0}^{\infty} A_m(x-\xi)^m,$$

where $A_m \in \mathscr{L}_s(^m E; F)$ $(m=0,1,\ldots)$; or again, of the form

$$\sum_{m=0}^{\infty} P_m(x-\xi),$$

where $P_m = \hat{A}_m \in \mathscr{P}(^m E; F)$ $(m=0,1,\ldots)$. Both the A_m and the P_m are called the *coefficients* of the power series.

Definition 2. The *radius of convergence* of a power series about ξ is the largest r, $0 \leqslant r \leqslant \infty$, such that the power series is uniformly convergent on every $\bar{B}_\rho(\xi)$ for $0 \leqslant \rho < r$. The power series is said to be *convergent* in case its radius of convergence is strictly positive; that is, if there is some $\rho > 0$ such that the power series converges uniformly on $\bar{B}_\rho(\xi)$.

Proposition 1 (Cauchy-Hadamard). *The radius of convergence of the power series is given by*

$$r = \frac{1}{\limsup\limits_{m \to \infty} \|P_m\|^{1/m}}.$$

Proof. Let us assume that the power series is uniformly convergent on a closed ball $\bar{B}_\rho(\xi)$. Call $f(x)$ its sum for $x \in \bar{B}_\rho(\xi)$.

There is some integer $M \geqslant 0$ such that

$$\left\| f(x) - \sum_{\ell=0}^{m} P_\ell(x-\xi) \right\| \leqslant 1$$

for $m \geqslant M$ and $x \in \bar{B}_\rho(\xi)$. Hence

$$\| P_m(x-\xi) \| \leqslant 2$$

for $m > M$ and $x \in \bar{B}_\rho(\xi)$. If $t \in E$, $\|t\| = 1$, we shall have

$$\| P_m(\rho t) \| = \rho^m \| P_m(t) \| \leqslant 2$$

for $m > M$. Hence

$$\rho^m \| P_m \| \leqslant 2$$

for $m > M$. We then get $\rho \leqslant 1/\Lambda$, where

$$\Lambda = \limsup_{m \to \infty} \| P_m \|^{1/m}.$$

It follows that $r \leqslant 1/\Lambda$. Let us now prove the opposite inequality $r \geqslant 1/\Lambda$. This is clear if $\Lambda = \infty$. Assume $\Lambda < \infty$. Take ρ such that $0 \leqslant \rho < 1/\Lambda$. Fix σ so that $\rho < \sigma < 1/\Lambda$. Hence $\sigma > 0$. Since $\Lambda < 1/\sigma$, there is some $C \geqslant 0$ such that $\| P_m \| \leqslant C/\sigma^m$ for $m = 0, 1, \ldots$ Hence

$$\| P_m(x-\xi) \| \leqslant \| P_m \| \cdot \| x - \xi \|^m \leqslant C \left(\frac{\rho}{\sigma} \right)^m$$

for $x \in \bar{B}_\rho(\xi)$ and $m = 0, 1, \ldots$ This proves that the power series is uniformly convergent on $\bar{B}_\rho(\xi)$. Therefore $\rho \leqslant r$. Letting $\rho \to 1/\Lambda$, we get $1/\Lambda \leqslant r$. This proves the Cauchy-Hadamard formula $r = 1/\Lambda$. *Q. E. D.*

Corollary 1. *The power series is convergent if and only if the sequence*

$$\| P_m \|^{1/m} \qquad (m = 1, 2, \ldots)$$

is bounded; or, equivalently, if and only if the sequence

$$\|A_m\|^{1/m} \qquad (m=1,2,\ldots)$$

is bounded.

Proof. Necessity and sufficiency of boundedness of the first sequence follows from Proposition 1. Now, boundedness of the first sequence is equivalent to boundedness of the second sequence in view of Proposition 1, § 3 and the classical remark that the sequence

$$\frac{m}{\sqrt[m]{m!}} \qquad (m=1,2,\ldots)$$

is bounded; as a matter of fact this sequence tends to the number e as a consequence of Stirling's formula

$$\lim_{m\to\infty} \frac{m!}{m^m e^{-m}\sqrt{2\pi m}} = 1. \qquad Q.E.D.$$

Remark 1. In the above Corollary 1, it makes no difference whether we use $\|P_m\|$ or $\|A_m\|$, where $P_m=\hat{A}_m$. It can be shown that this is also true of Proposition 1 itself, in case E is finite dimensional (see the comment at the end of Remark 1, § 3). This is no longer the case if E is infinite dimensional. The following is a simple example (compare with Remark 1, § 3). Let E be the Banach space ℓ^1 of all sequences $x=(x_1,\ldots,x_n,\ldots)$ of complex numbers such that

$$\|x\|=\sum|x_n|<\infty$$

normed accordingly. Take $F=\mathbb{C}$. Let $P_m\colon E\to F$ be defined by

$$P_m(x)=m^m\cdot(x_1\ldots x_m)$$

if $x=(x_1,\dots,x_n,\dots)\in E$ and $m=1,2,\dots,$ and $P_0=1$. Then $P_m\in\mathscr{P}(^mE;F)$ and it is easily seen that $\|P_m\|=1$ for $m=0,1,\dots$. Accordingly, the radius of convergence of the power series

$$\sum_{m=0}^{\infty} P_m(x)$$

is equal to 1, by the Cauchy-Hadamard formula. However, if $A_m\in\mathscr{L}_s(E;F)$ corresponds to P_m, it is easily seen that

$$\|A_m\| = \frac{m^m}{m!}$$

for $m=0,1,\dots$. Hence

$$\lim_{m\to\infty} \|P_m\|^{1/m}=1,$$

$$\lim_{m\to\infty} \|A_m\|^{1/m}=e,$$

by the Stirling formula.

Proposition 2. *If there is some* $\rho>0$ *such that the power series*

$$\sum_{m=0}^{\infty} P_m(x-\xi)$$

converges and has sum equal to zero, for every $x\in\bar{B}_\rho(\xi)$, *then* $P_m=0$ *for* $m=0,1,\dots$.

The proof will use the following lemma.

Lemma 1. *If* $u_m\in F$ $(m=0,1,\dots)$, $\delta>0$ *and the series*

$$\sum_{m=0}^{\infty} u_m\lambda^m$$

converges and has sum equal to 0 *for every* $\lambda\in\mathbb{C}$, $|\lambda|\leqslant\delta$, *then* $u_m=0$ *for* $m=0,1,\dots$.

Proof. Letting $\lambda = 0$, we get $u_0 = 0$. Assume that we have proved that $u_0 = \cdots = u_{h-1} = 0$ for some $h \geqslant 1$. Let us prove inductively that $u_h = 0$. Since $\sum u_m \delta^m$ converges, we have $u_m \delta^m \to 0$ as $m \to \infty$; hence

$$C = \sup_{m \geqslant 0} \|u_m\| \cdot \delta^m < \infty .$$

Now

$$u_h = - \sum_{m=h+1}^{\infty} u_m \lambda^{m-h}$$

for $\lambda \neq 0$, $|\lambda| \leqslant \delta$, Hence

$$\|u_h\| \leqslant \sum_{m=h+1}^{\infty} \|u_m\| \cdot |\lambda|^{m-h} \leqslant \frac{C}{\delta^h} \frac{|\lambda|}{\delta - |\lambda|},$$

provided $\lambda \neq 0$, $|\lambda| < \delta$. Letting $\lambda \to 0$, we get $u_h = 0$. *Q. E. D.*

Proof of Proposition 2. Let $t \in E$, $\lambda \in \mathbb{C}$, $x = \xi + \lambda t$. We have then

$$\sum_{m=0}^{\infty} P_m(t) \lambda^m = 0$$

provided $\|x - \xi\| = |\lambda| \cdot \|t\| \leqslant \rho$. By Lemma 1, we get $P_m(t) = 0$ for all $t \in E$, hence $P_m = 0$ $(m = 0, 1, \ldots)$. *Q. E. D.*

§ 5. Holomorphic Mappings

Definition 1. A mapping $f: U \to F$ is said to be *holomorphic on* U if, corresponding to every $\xi \in U$, there are a power series

$$\sum_{m=0}^{\infty} P_m(x-\xi)$$

from E to F about ξ and some $\rho > 0$ such that $B_\rho(\xi) \subset U$ and

$$f(x) = \sum_{m=0}^{\infty} P_m(x-\xi)$$

uniformly for $x \in B_\rho(\xi)$. The sequence P_m $(m=0, 1, \ldots)$ is then unique at every point ξ, by Proposition 2, § 4. This convergent power series is called the *Taylor series of* f *at* ξ and then we write

$$f(x) \cong \sum_{m=0}^{\infty} P_m(x-\xi)$$

to denote this relationship between f and the power series. We shall denote by $\mathscr{H}(U; F)$ the vector space of all holomorphic mappings from U to F, with respect to pointwise vector operations.

Definition 2. Let $f \in \mathscr{H}(U; F)$ and

$$f(x) \cong \sum_{m=0}^{\infty} P_m(x-\xi)$$

be its Taylor series at $\xi \in U$. Let $P_m \in \mathscr{P}({}^m E; F)$ correspond to $A_m \in \mathscr{L}_s({}^m E; F)$ by $P_m = \hat{A}_m$ $(m=0, 1, \ldots)$. We set the notations

$$d^m f(\xi) = m!\, A_m,$$
$$\hat{d}^m f(\xi) = m!\, P_m,$$

so that we have the differential mappings

$$d^m f : x \in U \mapsto d^m f(x) \in \mathscr{L}_s(^m E; F),$$
$$\hat{d}^m f : x \in U \mapsto \hat{d}^m f(x) \in \mathscr{P}(^m E; F),$$

and the differentiation operators

$$d^m : f \in \mathscr{H}(U; F) \mapsto d^m f \in \mathscr{H}(U; \mathscr{L}_s(^m E; F)),$$
$$\hat{d}^m : f \in \mathscr{H}(U; F) \mapsto \hat{d}^m f \in \mathscr{H}(U; \mathscr{P}(^m E; F)),$$

of order $m = 0, 1, \ldots$ (see Proposition 3, § 7). The *Taylor polynomial* $\tau_{m,f,\xi}$ of order m of f at ξ is defined by

$$\tau_{m,f,\xi}(x) = \sum_{\ell=0}^{m} \frac{1}{\ell!} d^\ell f(\xi)(x-\xi)^\ell = \sum_{\ell=0}^{m} \frac{1}{\ell!} \hat{d}^\ell f(\xi)(x-\xi),$$

where $x \in E$.

Remark 1. Definition 1 is known as the Weierstrass concept of a holomorphic mapping. According to the so-called Cauchy-Riemann definition, $f : U \to F$ is holomorphic on U if, corresponding to every $\xi \in U$, there is some necessarily unique $A \in \mathscr{L}(^1 E; F)$ such that

$$\lim_{h \to 0} \frac{\| f(\xi + h) - f(\xi) - A(h) \|}{\| h \|} = 0.$$

It is a theorem of Goursat that the Weierstrass and the Cauchy-Riemann definitions of holomorphy are equivalent. The proof goes as follows. One firstly adapts the classical proof of Goursat's theorem when $E = F = \mathbb{C}$ to take care of the case in which $E = \mathbb{C}$ and F is arbitrary. Next one passes from this case to general E and F by a simple reasoning.

Remark 2. It is clear that being holomorphic is a local property in the following sense. If $f : U \to F$ is holomorphic and

$V \subset U$ is open and non-void, then $f|V: V \to F$ is holomorphic. Moreover, if $U = \bigcup_{\lambda \in \Lambda} V_\lambda$, where $V_\lambda \subset U$ is open and non-void, and $f|V_\lambda: V_\lambda \to F$ is holomorphic for every $\lambda \in \Lambda$, then $f: U \to F$ is holomorphic

Proposition 1. *We have* $\mathscr{P}(E;F) \subset \mathscr{H}(E;F)$.

Proof. It is sufficient to prove that $\mathscr{P}(^m E;F) \subset \mathscr{H}(E;F)$ for $m = 0, 1, \ldots$. Let $A \in \mathscr{L}_s(^m E;F)$. We have Newton's binomial formula

$$A(x+y)^m = \sum_{k=0}^{m} \binom{m}{k} A\, x^{m-k} y^k,$$

where

$$A\, x^{m-k}\, y^k = A\,(\underbrace{x, \ldots, x}_{m-k},\ \underbrace{y, \ldots, y}_{k})$$

for $x, y \in E$. It follows that

$$A\, x^m = \sum_{k=0}^{m} \binom{m}{k} A\, \xi^{m-k} (x-\xi)^k$$

where $\xi, x \in E$. Denote by $A\,\xi^{m-k}$ the element of $\mathscr{L}_s(^k E;F)$ given by

$$A\,\xi^{m-k}(y_1, \ldots, y_k) = A(\underbrace{\xi, \ldots, \xi}_{m-k}, y_1, \ldots, y_k).$$

Then we deduce that $P = \hat{A}$ is holomorphic on E and that actually

$$\frac{1}{k!}\, d^k P(\xi) = \binom{m}{k} A\, \xi^{m-k}$$

for $k = 0, \ldots, m$, and

$$d^k P(\xi) = 0$$

for $k > m$. *Q. E. D.*

Remark 3. From the above proof, we see, in particular, that

$$d^k P \in \mathscr{P}({}^{m-k}E; \mathscr{L}_s({}^k E; F)),$$
$$\hat{d}^k P \in \mathscr{P}({}^{m-k}E; \mathscr{P}({}^k E; F)),$$

for $P \in \mathscr{P}({}^m E; F)$ and $k = 0, \ldots, m$.

§ 6. The Cauchy Integral

Proposition 1 (Cauchy integral). *Let* $f \in \mathscr{H}(U; F)$, $\xi \in U$, $x \in U$ *and* $\rho > 1$ *be such that* $(1-\lambda)\xi + \lambda x \in U$ *for every* $\lambda \in \mathbb{C}$, $|\lambda| \leqslant \rho$. *Then*

$$f(x) = \frac{1}{2\pi i} \int_{|\lambda| = \rho} \frac{f[(1-\lambda)\xi + \lambda x]}{\lambda - 1} d\lambda.$$

Proof. First of all, let $V \subset \mathbb{C}$ be open, $g: V \to F$ be holomorphic and $\tau \in B_\delta(\zeta) \subset \bar{B}_\delta(\zeta) \subset V$ for some $\delta > 0$. Then

$$g(\tau) = \frac{1}{2\pi i} \int_{|\lambda - \zeta| = \delta} \frac{g(\lambda)}{\lambda - \tau} d\lambda.$$

This is proved for general F exactly as in the classical case $F = \mathbb{C}$. Let us apply such a fact to the proof of the proposition. Set

$$g(\lambda) = f[(1-\lambda)\xi + \lambda x],$$

where g is defined and easily seen to be holomorphic in the open subset V of $\mathbb{C}$ of all $\lambda \in \mathbb{C}$ such that $(1-\lambda)\xi + \lambda x \in U$. By the assumption, the closed disc in $\mathbb{C}$ of center at 0 and radius ρ is contained in V and contains 1 in its interior. Hence

$$g(1) = \frac{1}{2\pi i} \int_{|\lambda| = \rho} \frac{g(\lambda)}{\lambda - 1} d\lambda,$$

which is precisely what we want. *Q. E. D.*

Proposition 2 (Cauchy integral). *Let* $f \in \mathscr{H}(U; F)$, $\xi \in U$, $x \in E$ *and* $\rho > 0$ *be such that* $\xi + \lambda x \in U$ *for every* $\lambda \in \mathbb{C}$, $|\lambda| \leqslant \rho$. *Then*

$$\frac{1}{m!} \hat{d}^m f(\xi)(x) = \frac{1}{2\pi i} \int_{|\lambda| = \rho} \frac{f(\xi + \lambda x)}{\lambda^{m+1}} d\lambda$$

for $m = 0, 1, \ldots$.

Proof. First of all, let $V \subset \mathbb{C}$ be open, $g: V \to F$ be holomorphic, $0 < r \leqslant R$, $\xi \in \mathbb{C}$ and the closed annulus in $\mathbb{C}$ of center at ξ and radii r and R be contained in V.

Then

$$\int_{|\lambda - \xi| = R} g(\lambda)\, d\lambda = \int_{|\lambda - \xi| = r} g(\lambda)\, d\lambda.$$

This is proved for general F exactly as in the classical case $F = \mathbb{C}$. Let us use this fact in proving the proposition. Set

$$g(\lambda) = \frac{f(\xi + \lambda x)}{\lambda^{m+1}},$$

where g is defined and easily seen to be holomorphic in the open subset V of $\mathbb{C}$ of all $\lambda \in \mathbb{C}$ such that $\xi + \lambda x \in U$ and $\lambda \neq 0$. By the assumption, the closed disc in $\mathbb{C}$ of center at 0 and radius ρ is contained in V, except for its center 0. Hence, letting $0 < \varepsilon \leqslant \rho$,

$$\int_{|\lambda| = \rho} g(\lambda)\, d\lambda = \int_{|\lambda| = \varepsilon} g(\lambda)\, d\lambda,$$

that is,

$$\int_{|\lambda| = \rho} \frac{f(\xi + \lambda x)}{\lambda^{m+1}} d\lambda = \int_{|\lambda| = \varepsilon} \frac{f(\xi + \lambda x)}{\lambda^{m+1}} d\lambda.$$

Consider the Taylor series

$$f(t) \cong \sum_{\ell = 0}^{\infty} P_{\ell}(t - \xi)$$

of f at ξ, and assume ε small enough so that this power series will converge to $f(t)$ uniformly for t in the closed ball in E of center at ξ and radius $\varepsilon \cdot \|x\|$. We then get

$$\int_{|\lambda|=\rho} \frac{f(\xi+\lambda x)}{\lambda^{m+1}} d\lambda = \sum_{\ell=0}^{\infty} P_\ell(x) \int_{|\lambda|=\varepsilon} \frac{d\lambda}{\lambda^{m+1-\ell}} = 2\pi i P_m(x),$$

from which we get the proposition. *Q. E. D.*

Proposition 3 (Cauchy inequalities). *Let* $f \in \mathscr{H}(U;F)$, $\rho > 0$ *and* $\bar{B}_\rho(\xi) \subset U$. *Then*

$$\left\| \frac{1}{m!} \hat{d}^m f(\xi) \right\| \leqslant \frac{1}{\rho^m} \sup_{\|x-\xi\|=\rho} \|f(x)\|$$

for $m = 0, 1, \ldots$.

Proof. Apply Proposition 2 by assuming $\|x\| = 1$. *Q. E. D.*

Remark 1. Let $\xi \in E$, $\rho > 0$, and $C \geqslant 0$ be given so that

$$\left\| \frac{1}{m!} \hat{d}^m f(\xi) \right\| \leqslant C \cdot \sup_{\|x-\xi\|=\rho} \|f(x)\|$$

holds for every $f \in \mathscr{H}(E;F)$. Take $P \in \mathscr{P}(^m E;F)$, $f(x) = P(x-\xi)$ for $x \in E$. Then

$$\frac{1}{m!} \hat{d}^m f(x) = P$$

for every $x \in E$, and we get

$$\|P\| \leqslant C \rho^m \|P\|,$$

that is $C \geqslant 1/\rho^m$ by taking $P \neq 0$. This shows that $1/\rho^m$ is the best universal coefficient for Proposition 3. On the other hand,

Proposition 3 would also hold in the form

$$\left\| \frac{1}{m!} d^m f(\xi) \right\| \leqslant \frac{1}{\rho^m} \frac{m^m}{m!} \sup_{\|x-\xi\|=\rho} \|f(x)\|$$

by Proposition 3 itself and Proposition 1, § 3. Moreover, the above reasoning would show that

$$\frac{1}{\rho^m} \frac{m^m}{m!}$$

is then the best universal coefficient, by Remark 1, § 3.

Proposition 4. *$\mathscr{H}(U;F)$ is contained and closed in $\mathscr{C}(U;F)$, the latter space being endowed with the compact-open topology.*

The proof will be based on the following lemma.

Lemma 1. *Let $f \in \mathscr{H}(U;F)$, $\xi \in U$, $x \in U$ and the real number $\rho > 1$ be such that $(1-\lambda)\xi + \lambda x \in U$ for every $\lambda \in \mathbb{C}$, $|\lambda| \leqslant \rho$. Then*

$$\|f(x) - \tau_{m,f,\xi}(x)\| \leqslant \frac{1}{\rho^m(\rho-1)} \sup_{|\lambda|=\rho} \|f[(1-\lambda)\xi + \lambda x]\|$$

for $m = 0, 1, \ldots$.

Proof. By Propositions 1 and 2 and the identity

$$\frac{1}{\lambda-1} = \sum_{k=0}^{m} \frac{1}{\lambda^{k+1}} + \frac{1}{\lambda^{m+1}(\lambda-1)}$$

for $\lambda \neq 0$ and $\lambda \neq 1$, we get

$$f(x) - \tau_{m,f,\xi}(x) = \frac{1}{2\pi i} \int_{|\lambda|=\rho} \frac{f[(1-\lambda)\xi + \lambda x]}{\lambda^{m+1}(\lambda-1)} d\lambda,$$

from which the lemma follows. *Q. E. D.*

Proof of Proposition 4. Let $\xi \in U$ be given. Choose $\rho > 0$ so that $B_\rho(\xi) \subset U$ and, letting $f \in \mathscr{H}(U;F)$, we have

$$f(x) = \sum_{m=0}^{\infty} P_m(x-\xi)$$

uniformly for $x \in B_\rho(\xi)$, where the series in question is the Taylor series of f at ξ. By Corollary 1, § 4, the sequence

$$\|P_m\|^{1/m} \qquad (m=1,2,\dots)$$

is bounded by some $c \geqslant 0$. Hence

$$\|f(x)-f(\xi)\| \leqslant \sum_{m=1}^{\infty} \|P_m\| \cdot \|x-\xi\|^m \leqslant \frac{c\,\|x-\xi\|}{1-c\,\|x-\xi\|}$$

provided $\|x-\xi\| < \rho$ if we further assume $\rho c \leqslant 1$. It follows that f is continuous at ξ. Hence

$$\mathscr{H}(U;F) \subset \mathscr{C}(U;F).$$

Let us now prove that $\mathscr{H}(U;F)$ is closed in $\mathscr{C}(U;F)$ with respect to the compact-open topology. Let f belong to the closure $\overline{\mathscr{H}(U;F)}$ of $\mathscr{H}(U;F)$ in $\mathscr{C}(U;F)$. Fix $\xi \in U$. Given any $x \in E$, choose $\rho > 0$ so that the closed ball in E of center at ξ and radius $\rho \cdot \|x\|$ be contained in U. Define $\hat{d}^m f(\xi): E \to F$ by

$$\frac{1}{m!}\hat{d}^m f(\xi)(x) = \frac{1}{2\pi i} \int_{|\lambda|=\rho} \frac{f(\xi+\lambda x)}{\lambda^{m+1}}\, d\lambda.$$

Notice that the right-hand side of this equality does not depend on the choice of such ρ; this is indeed the case if $f \in \mathscr{H}(U;F)$ (see the proof of Proposition 2), hence if $f \in \overline{\mathscr{H}(U;F)}$. We have obviously

$$\hat{d}^m f(\xi)(\mu x) = \mu^m \hat{d}^m f(\xi)(x)$$

for $\mu \in \mathbb{C}$, $x \in E$. Use the polarization formula (see the proof of Proposition 1, § 3) to define $d^m f(\xi): E^m \to F$ by

$$d^m f(\xi)(x_1, \dots, x_m) = \frac{1}{m!\,2^m} \sum_{\substack{\varepsilon_1 = \pm 1 \\ \cdots\cdots \\ \varepsilon_m = \pm 1}} \varepsilon_1 \dots \varepsilon_m \hat{d}^m f(\xi)(\varepsilon_1 x_1 + \cdots + \varepsilon_m x_m)$$

for $x_1, \dots, x_m \in E$. Then $d^m f(\xi)$ is m-linear and symmetric; this is indeed the case if $f \in \mathscr{H}(U;F)$, hence if $f \in \overline{\mathscr{H}(U;F)}$. It is clear that

$$\hat{d}^m f(\xi)(x) = d^m f(\xi)\, x^m$$

for $x \in E$, so that $\hat{d}^m f(\xi)$ is the m-homogeneous polynomial associated to $d^m f(\xi)$. We claim that $\hat{d}^m f(\xi)$ is continuous. In fact, if $\|x\| = 1$ and ρ is such that $\bar{B}_\rho(\xi) \subset U$ and f is bounded on $\bar{B}_\rho(\xi)$, we get

$$\left\| \frac{1}{m!} \hat{d}^m f(\xi)(x) \right\| \leqslant \frac{1}{\rho^m} \sup_{\|t - \xi\| = \rho} \|f(t)\| < \infty .$$

It follows that $\hat{d}^m f(\xi)$ is continuous, hence $d^m f(\xi)$ is continuous too. We thus have

$$d^m f(\xi) \in \mathscr{L}_s({}^m E; F),$$
$$\hat{d}^m f(\xi) \in \mathscr{P}({}^m E; F).$$

Define $\tau_{m,f,\xi}$ for f, exactly as in Definition 2, § 5. The estimate in Lemma 1 holds for $f \in \overline{\mathscr{H}(U;F)}$, since Proposition 1 holds for $f \in \overline{\mathscr{H}(U;F)}$. Choose $r > 0$ and $\rho > 1$ such that $B_{\rho r}(\xi) \subset U$ and f is bounded in $B_{\rho r}(\xi)$. We then conclude that f is the limit of $\tau_{m,f,\xi}$ as $m \to \infty$ uniformly on $B_r(\xi)$. Hence $f \in \mathscr{H}(U;F)$. *Q. E. D.*

§ 7. Convergence of Taylor Series

Proposition 1. *Let $f \in \mathscr{H}(U;F)$, $\xi \in U$ and U be ξ-equilibrated. Then the Taylor series of f at ξ converges to f uniformly on some neighborhood contained in U of every compact subset of U.*

Proof. It is sufficient to prove that the Taylor series of f at ξ converges to f uniformly on some neighborhood $V \subset U$ of every $x \in U$. In fact, once $x \in U$ is given, choose $\rho > 1$ and a neighborhood $V \subset U$ of x such that

$$\sup_{t \in V, |\lambda| = \rho} \| f[(1-\lambda)\xi + \lambda t] \| < \infty,$$

and apply Lemma 1, § 6. *Q. E. D.*

Definition 1. If $f \in \mathscr{H}(U;F)$ and $\xi \in U$, the *radius of boundedness* of f at ξ is the largest r, $0 < r \leqslant \infty$, such that $B_r(\xi) \subset U$ and f is bounded on every $\bar{B}_\rho(\xi)$, $0 \leqslant \rho < r$.

Proposition 2. *Let $f \in \mathscr{H}(U;F)$ and $\xi \in U$. The radius of boundedness r_b of f at ξ is the infimum of the radius of convergence r_c of the Taylor series of f at ξ and the distance d of ξ to the boundary of U.*

Proof. Let us start by proving $r_b \leqslant \inf(r_c, d)$.

In fact, $r_b \leqslant d$ is clear. Consider next $r_b \leqslant r_c$. Let

$$f(x) \cong \sum_{m=0}^{\infty} P_m(x-\xi)$$

be the Taylor series of f at ξ. If $0<\rho<r_b$ and

$$M = \sup_{\|x-\xi\| \leqslant \rho} \|f(x)\| < \infty$$

then

$$\|P_m\| \leqslant \frac{M}{\rho^m} \qquad (m=0,1,\ldots)$$

by the Cauchy inequalities. Hence

$$\lim_{m\to\infty} \sup \|P_m\|^{1/m} \leqslant \frac{1}{\rho}$$

and so $\rho \leqslant r_c$ by the Cauchy-Hadamard formula. Letting $\rho \to r_b$, we get $r_b \leqslant r_c$.

Let us now prove

$$r_b \geqslant \inf(r_c, d).$$

For every ρ, $0<\rho<\inf(r_c,d)$, we have that the Taylor series of f at ξ converges uniformly on $\bar{B}_\rho(\xi)$, since $\rho<r_c$, and has there sum equal to f, by Proposition 1 and $\rho<d$. Now, every continuous polynomial is bounded on $\bar{B}_\rho(\xi)$. Hence f is bounded on $\bar{B}_\rho(\xi)$ for every such ρ. Letting $\rho \to \inf(r_c,d)$, we obtain the desired inequality. *Q. E. D.*

Remark 1. If E is finite dimensional, every closed ball of finite radius in E is compact, and every $f \in \mathscr{H}(U;F) \subset \mathscr{C}(U;F)$ is bounded on each compact subset of U. It follows that the radius of boundedness r_b of f at ξ is equal to the distance d of ξ to the boundary of U. The above proposition expresses then that the radius of convergence r_c of the Taylor series of f at ξ is at least equal to d. In other words, we have

$$r_b = d \leqslant r_c$$

in the finite dimensional case. In case, however, E is infinite dimensional, it may occur that

$$r_b = r_c < d.$$

The following is a simple example. Let E be the Banach space c_0 of all sequences $x = (x_1, \ldots, x_n, \ldots)$ of complex numbers tending to 0, normed by

$$\|x\| = \sup_{n \geqslant 1} |x_n|.$$

Let $P_m \in \mathscr{P}(^m E; \mathbb{C})$ be defined by

$$P_m(x) = x_1 \ldots x_m$$

if $x = (x_1, \ldots, x_n, \ldots) \in E$ and $m = 1, 2, \ldots$, and $P_0 = 1$. Consider the Taylor series

$$\sum_{m=0}^{\infty} P_m(x)$$

for $x \in E$. It is easily seen that this Taylor series converges for every $x \in E$; and that, actually, if $0 \leqslant \rho < 1$, it converges uniformly for $x \in \bar{B}_\rho(t)$ for every $t \in E$. It follows that the function $f: E \to \mathbb{C}$ defined by this series is holomorphic on E, by Proposition 4, § 6 (or by direct proof). In this case, $\xi = 0$ and $d = \infty$. On the other hand, it is easily seen that $\|P_m\| = 1$ $(m = 0, 1, \ldots)$. Hence $r_c = 1$ by the Cauchy-Hadamard formula, and $r_b = \inf(r_c, d) = 1$ by Proposition 2.

Thus

$$r_b = 1, \qquad r_c = 1, \qquad d = \infty.$$

As a matter of fact, we could have seen that $r_b = 1$ as follows. Notice that

$$|f(x)| \leqslant \sum_{m=0}^{\infty} |P_m(x)| \leqslant \sum_{m=0}^{\infty} \|x\|^m = \frac{1}{1 - \|x\|}$$

if $\|x\|<1$. Hence $r_b \geqslant 1$. On the other hand, if x denotes the sequence

$$(\underbrace{1,\ldots,1}_{n},0,0,\ldots)\in E$$

we see that $\|x\|=1$ and $f(x)=n+1$. Hence f is unbounded on the closed ball in E of center at 0 and radius 1. This shows that $r_b \leqslant 1$. We conclude once again that $r_b=1$, whereas $d=\infty$.

Proposition 3. *If* $f\in\mathscr{H}(U;F)$, *then*

$$d^m f\in\mathscr{H}(U;\mathscr{L}_s(^mE;F)),$$
$$\hat{d}^m f\in\mathscr{H}(U;\mathscr{P}(^mE;F)),$$

for $m=0,1,\ldots$. *If*

$$f(x)\cong\sum_{k=0}^{\infty}P_k(x-\xi)$$

is the Taylor series of f *at* $\xi\in U$, *then the Taylor series of* $d^m f$ *and* $\hat{d}^m f$ *at* ξ *are*

$$d^m f(x)\cong\sum_{k=0}^{\infty}d^m P_{k+m}(x-\xi),$$
$$\hat{d}^m f(x)\cong\sum_{k=0}^{\infty}\hat{d}^m P_{k+m}(x-\xi).$$

Proof. Let r_b be the radius of boundedness of f at ξ. By Propositions 1 and 2, the Taylor series of f at ξ converges to $f(x)$ uniformly for $x\in\bar{B}_\rho(\xi)$, for any $0\leqslant\rho<r_b$. By the Cauchy inequalities, the prospective Taylor series of $\hat{d}^m f$ at ξ converges to $\hat{d}^m f(x)$ uniformly for $x\in\bar{B}_\rho(\xi)$ too, since we can increase ρ in the previous argument. We next notice that

$$\hat{d}^m P_{k+m}\in\mathscr{P}(^kE;\mathscr{P}(^mE;F))$$

to conclude the proof of the assertions concerning $\hat{d}^m f$. The corresponding assertions for $d^m f$ then follows from there. *Q. E. D.*

Corollary 1. *If* $f \in \mathscr{H}(U;F)$ *and* $\xi \in U$, *then, for* $k, m = 0, 1, \ldots,$

$$\frac{1}{k!}\hat{d}^k\left(\frac{1}{m!}\hat{d}^m f\right)(\xi) = \frac{1}{m!}\hat{d}^m\left[\frac{1}{(k+m)!}\hat{d}^{k+m} f(\xi)\right].$$

Proof. In the notation of the proposition, we have

$$P_k = \frac{1}{k!}\hat{d}^k f(\xi),$$

$$\hat{d}^m P_{k+m} = \frac{1}{k!}\hat{d}^k(\hat{d}^m f)(\xi),$$

from which the Corollary follows. *Q. E. D.*

Corollary 2. *If* $f \in \mathscr{H}(U;F)$ *and* $\xi \in U$, *then, for* $k, m = 0, 1, \ldots,$

$$\tau_{k, \hat{d}^m f, \xi} = \hat{d}^m(\tau_{k+m, f, \xi}).$$

Proof. Apply the Proposition or Corollary 1. *Q. E. D.*

§ 8. Topology on the Space of all Holomorphic Mappings

In the present section, we shall summarize the definition and main properties of the topology $\mathscr{I}_\omega$ on $\mathscr{H}(U;F)$. No proofs will be given here. Proofs of the major facts will be presented in the more general case of the topology $\mathscr{I}_{\omega,\Theta}$ on $\mathscr{H}_\Theta(U;F)$, starting from § 9 on. We shall indicate in § 14 how the general considerations developed in the bulk of this monograph, § 9 to § 13, simplify in the case of $\mathscr{H}(U;F)$.

A seminorm p on $\mathscr{H}(U;F)$ is said to be *ported* by a compact subset K of U if to every open subset V of U containing K there corresponds a real number $c(V)>0$ such that

$$p(f)\leqslant c(V)\cdot \sup_{x\in V} \|f(x)\|$$

for every $f\in\mathscr{H}(U;F)$. The topology $\mathscr{I}_\omega$ on $\mathscr{H}(U;F)$ is defined by all such seminorms ported by compact subsets of U. Each of the following conditions is necessary and sufficient for p to be ported by K:

(1) Corresponding to every real number $\varepsilon>0$ there is a real number $c(\varepsilon)>0$ such that, for every $f\in\mathscr{H}(U;F)$,

$$p(f)\leqslant c(\varepsilon)\sum_{m=0}^{\infty} \varepsilon^m \sup_{x\in K} \left\| \frac{1}{m!}\hat{d}^m f(x)\right\|.$$

(2) Corresponding to every real number $\varepsilon>0$ and open subset V of U containing K, there is a real number $c(\varepsilon, V)>0$

such that, for every $f \in \mathscr{H}(U;F)$,

$$p(f) \leqslant c(\varepsilon, V) \sum_{m=0}^{\infty} \varepsilon^m \sup_{x \in V} \left\| \frac{1}{m!} \hat{d}^m f(x) \right\| .$$

If U is ξ-equilibrated, the Taylor series at ξ of any $f \in \mathscr{H}(U;F)$ converges to f in the sense of $\mathscr{I}_\omega$. If K is ξ-equilibrated, the following condition is necessary and sufficient for p to be ported by K: corresponding to every open subset V of U containing K, there is a real number $c(V) > 0$ such that, for every $f \in \mathscr{H}(U;F)$,

$$p(f) \leqslant c(V) \cdot \sum_{m=0}^{\infty} \sup_{x \in V} \left\| \frac{1}{m!} \hat{d}^m f(\xi)(x-\xi) \right\| .$$

The compact-open topology on the vector space $\mathscr{C}(U;F)$ of all continuous F-valued functions on U induces a topology $\mathscr{I}_0$ on $\mathscr{H}(U;F)$. We have $\mathscr{I}_0 \subset \mathscr{I}_\omega$; $\mathscr{I}_0 = \mathscr{I}_\omega$ if and only if $\dim E < \infty$, or $F = 0$. Each $\hat{d}^m$ is continuous for the corresponding topologies $\mathscr{I}_\omega$; continuity of $\hat{d}^m$ for some $m \geqslant 1$ and the corresponding topologies $\mathscr{I}_0$ requires $\dim E < \infty$, or $F = 0$. However a subset $\mathscr{X}$ of $\mathscr{H}(U;F)$ is bounded for $\mathscr{I}_\omega$ if and only if it is bounded for $\mathscr{I}_0$. Each of the following conditions is necessary and sufficient for $\mathscr{X}$ to be bounded for $\mathscr{I}_\omega$:

(1) Corresponding to every compact subset K of U, there is a real number $C \geqslant 0$ such that $\|f(x)\| \leqslant C$ for every $f \in \mathscr{X}$ and $x \in K$.

(2) Corresponding to every compact subset K of U, there are a real number $C \geqslant 0$ and an open subset V of U containing K such that $\|f(x)\| \leqslant C$ for every $f \in \mathscr{X}$ and $x \in V$.

(1′) Corresponding to every $\xi \in U$, there are real numbers $C \geqslant 0$ and $c \geqslant 0$ such that, for every $m = 0, 1, \ldots$ and $f \in \mathscr{X}$,

$$\left\| \frac{1}{m!} \hat{d}^m f(\xi) \right\| \leqslant C \cdot c^m .$$

(2′) Corresponding to every compact subset K of U, there are real numbers $C \geqslant 0$ and $c \geqslant 0$ such that, for every $m=0,1,\ldots, f \in \mathscr{X}$ and $x \in K$,

$$\left\| \frac{1}{m!} \hat{d}^m f(x) \right\| \leqslant C \cdot c^m .$$

(3′) Corresponding to every compact subset K of U, there are real numbers $C \geqslant 0$ and $c \geqslant 0$, and an open subset V of U containing K, such that, for every $m=0,1,\ldots, f \in \mathscr{X}$ and $x \in V$,

$$\left\| \frac{1}{m!} \hat{d}^m f(x) \right\| \leqslant C \cdot c^m .$$

Let $X \subset U$ be fixed, and suppose X meets every connected component of U. Then $\mathscr{X}$ is bounded for $\mathscr{I}_\omega$ if and only if $\mathscr{X}$ is equicontinuous on U and $\sup\{\|f(x)\| \mid f \in \mathscr{X}\} < \infty$ for every $x \in X$. Denote by $\mathscr{I}_{\infty,X}$ the topology on $\mathscr{H}(U;F)$ defined by the family of seminorms $f \to \|\hat{d}^m f(x)\|$ for $m=0,1,\ldots$ and $x \in X$. If $\mathscr{X}$ is $\mathscr{I}_\omega$-bounded, $\mathscr{I}_\omega$ and $\mathscr{I}_{\infty,X}$ induce the same topology on $\mathscr{X}$; also the uniform structures associated with $\mathscr{I}_\omega$ and $\mathscr{I}_{\infty,X}$ induce the same uniform structure on $\mathscr{X}$. If f, $f_\ell \in \mathscr{H}(U;F)$ for $\ell=0,1,\ldots$, then $f_\ell \to f$ in the sense of $\mathscr{I}_\omega$ as $\ell \to \infty$ if and only if $\{f_\ell\}$ is $\mathscr{I}_\omega$-bounded and $\hat{d}^m f_\ell(x) \to \hat{d}^m f(x)$ in $\mathscr{P}(^m E;F)$ as $\ell \to \infty$ for every $m=0,1,\ldots$ and $x \in X$. Also $\mathscr{X}$ is $\mathscr{I}_\omega$-relatively compact if and only if $\mathscr{X}$ is $\mathscr{I}_\omega$-bounded and $\{\hat{d}^m f(x) \mid f \in \mathscr{X}\}$ is relatively compact in $\mathscr{P}(^m E;F)$ for every $m=0,1,\ldots$ and $x \in X$.

§ 9. Holomorphy Types

Definition 1. A *holomorphy type* Θ from E to F is a sequence of Banach spaces $\mathscr{P}_\Theta(^m E; F)$, for $m \in \mathbb{N}$, the norm on each of them being denoted by $P \to \|P\|_\Theta$, such that the following conditions hold true:

(1) Each $\mathscr{P}_\Theta(^m E; F)$ is a vector subspace of $\mathscr{P}(^m E; F)$.

(2) $\mathscr{P}_\Theta(^0 E; F)$ coincides with $\mathscr{P}(^0 E; F)$ as a normed vector space.

(3) There is a real number $\sigma \geqslant 1$ for which the following is true. Given any $\ell \in \mathbb{N}$, $m \in \mathbb{N}$, $\ell \leqslant m$, $x \in E$, and $P \in \mathscr{P}_\Theta(^m E; F)$, we have

$$\hat{d}^\ell P(x) \in \mathscr{P}_\Theta(^\ell E; F),$$

$$\left\| \frac{1}{\ell!} \hat{d}^\ell P(x) \right\|_\Theta \leqslant \sigma^m \cdot \|P\|_\Theta \cdot \|x\|^{m-\ell}.$$

Proposition 1. *Each inclusion mapping*

$$\mathscr{P}_\Theta(^m E; F) \subset \mathscr{P}(^m E; F)$$

is continuous and of norm inferior to σ^m, $m \in \mathbb{N}$.

Proof. Set $\ell = 0$ in Condition (3) and use Conditions (1) and (2) of Definition 1 to get

$$\|P(x)\| \leqslant \sigma^m \cdot \|P\|_\Theta \cdot \|x\|^m$$

that is

$$\|P\| \leqslant \sigma^m \|P\|_\Theta . \qquad \text{Q. E. D.}$$

Definition 2. A given $f \in \mathscr{H}(U;F)$ is said to be of *Θ-holomorphy type at* $\xi \in U$ if:

(1) $\hat{d}^m f(\xi) \in \mathscr{P}_\Theta({}^m E; F)$ for $m \in \mathbb{N}$.

(2) There are real numbers $C \geqslant 0$ and $c \geqslant 0$ such that

$$\left\| \frac{1}{m!} \hat{d}^m f(\xi) \right\|_\Theta \leqslant C \cdot c^m \quad \text{for} \quad m \in \mathbb{N}.$$

Moreover, f is said to be of *Θ-holomorphy type on* U if f is of Θ-holomorphy type at every point of U. We shall denote by $\mathscr{H}_\Theta(U;F)$ the vector subspace of $\mathscr{H}(U;F)$ of all such f of Θ-holomorphy type on U.

Proposition 2. *If $f \in \mathscr{H}_\Theta(U;F)$ then corresponding to every compact subset K of U there are real numbers $C \geqslant 0$ and $c \geqslant 0$, and an open subset V of U containing K, such that*

$$\left\| \frac{1}{m!} \hat{d}^m f(x) \right\|_\Theta \leqslant C \cdot c^m$$

for every $x \in V$ and $m \in \mathbb{N}$.

The proof will be based on the following lemma.

Lemma 1. *Let $f \in \mathscr{H}(U;F)$ be of Θ-holomorphy type at $\xi \in U$. Then there exists a real number $\rho > 0$ such that $B_\rho(\xi) \subset U$, and such that:*

(1) *$\hat{d}^m f(x) \in \mathscr{P}_\Theta({}^m E; F)$ for every $x \in B_\rho(\xi)$ and $m \in \mathbb{N}$.*

(2) *There are real numbers $C \geqslant 0$ and $c \geqslant 0$ such that*

$$\left\| \frac{1}{m!} \hat{d}^m f(x) \right\|_\Theta \leqslant C \cdot c^m$$

for every $x \in B_\rho(\xi)$ and $m \in \mathbb{N}$.

(3) *For every* $x \in B_\rho(\xi)$ *and* $\ell \in \mathbb{N}$,

$$\hat{d}^\ell f(x) = \sum_{m=\ell}^{\infty} \hat{d}^\ell P_m(x-\xi),$$

where

$$P_m = \frac{1}{m!} \hat{d}^m f(\xi) \qquad (m \in \mathbb{N}),$$

convergence of the series being in the sense of $\mathscr{P}_\Theta({}^\ell E; F)$.

Proof. Consider the Taylor series

$$f(x) \cong \sum_{m=0}^{\infty} P_m(x-\xi)$$

of f at ξ. Then there exists some real number $\rho > 0$ (take $0 < \rho < r_b$; see the proof of Proposition 3, § 7) such that $B_\rho(\xi) \subset U$, and such that, letting $x \in B_\rho(\xi)$ and $\ell \in \mathbb{N}$, we have

$$(*) \qquad \hat{d}^\ell f(x) = \sum_{m=\ell}^{\infty} \hat{d}^\ell P_m(x-\xi),$$

convergence of this series being in the sense of $\mathscr{P}({}^\ell E; F)$. By Condition (1) of Definition 2, we have $P_m \in \mathscr{P}_\Theta({}^m E; F)$. Hence $\hat{d}^\ell P_m(x-\xi) \in \mathscr{P}_\Theta({}^\ell E; F)$, by Condition (3) of Definition 1. Using Condition (3) of Definition 1, we have

$$\sum_{m=\ell}^{\infty} \|\hat{d}^\ell P_m(x-\xi)\|_\Theta \leqslant \ell! \sum_{m=\ell}^{\infty} \sigma^m \cdot \|P_m\|_\Theta \cdot \|x-\xi\|^{m-\ell}.$$

There are real numbers $C \geqslant 0$ and $c \geqslant 0$ such that $\|P_m\|_\Theta \leqslant C \cdot c^m$ for every $m \in \mathbb{N}$, by Condition (2) of Definition 2. If, in addition, we assume that ρ was chosen sufficiently small so that $\sigma c \rho < 1$, we shall have

$$\sum_{m=\ell}^{\infty} \|\hat{d}^\ell P_m(x-\xi)\|_\Theta \leqslant \frac{\ell! C}{1-\sigma c \rho} \cdot (\sigma c)^\ell$$

for every $x \in B_\rho(\xi)$ and $\ell \in \mathbb{N}$, so that the series in (*) will converge in the sense of $\mathcal{P}_\Theta({}^\ell E; F)$, by completeness of this normed space. Since that series converges to $\hat{d}^\ell f(x)$ in the sense of $\mathcal{P}({}^\ell E; F)$, by assumption, it follows, by virtue of Proposition 1, that actually $\hat{d}^\ell f(x) \in \mathcal{P}_\Theta({}^\ell E; F)$, that (*) is true in the sense of $\mathcal{P}_\Theta({}^\ell E; F)$, and also that

$$\left\| \frac{1}{\ell!} \hat{d}^\ell f(x) \right\|_\Theta \leqslant \frac{C}{1-\sigma c \rho} \cdot (\sigma c)^\ell$$

for every $x \in B_\rho(\xi)$ and $\ell \in \mathbb{N}$. *Q. E. D.*

Proof of Proposition 2. Apply Condition (2) of Lemma 1.
Q. E. D.

§ 10. Differentiation of Holomorphy Types

Definition 1. Once a holomorphy type Θ from E to F and $\ell \in \mathbb{N}$ are given, the vector space isomorphism

$$P \in \mathscr{P}(^{\ell+m}E; F) \mapsto \frac{1}{\ell!} \hat{d}^\ell P \in \mathscr{P}(^m E; \mathscr{P}(^\ell E; F))$$

(for each $m \in \mathbb{N}$) induces, by Condition (3) of Definition 1, § 9, a vector space isomorphism of $\mathscr{P}_\Theta(^{\ell+m}E; F)$ onto a vector subspace of $\mathscr{P}(^m E; \mathscr{P}_\Theta(^\ell E; F))$. Such a vector subspace will be denoted by

$$\frac{1}{\ell!} \hat{d}^\ell \mathscr{P}_\Theta(^{\ell+m}E; F);$$

it becomes a Banach space when normed in an isometric way by

$$\left\| \frac{1}{\ell!} \hat{d}^\ell P \right\|_\tau = \| P \|_\Theta$$

for $P \in \mathscr{P}_\Theta(^{\ell+m}E; F)$.

Proposition 1. *For each fixed holomorphy type Θ from E to F and $\ell \in \mathbb{N}$, the sequence of Banach spaces*

$$\frac{1}{\ell!} \hat{d}^\ell \mathscr{P}_\Theta(^{\ell+m}E; F) \qquad (m \in \mathbb{N})$$

is a holomorphy type τ from E to $\mathscr{P}_\Theta(^\ell E; F)$ (to be denoted by $\left.\frac{1}{\ell!} \hat{d}^\ell \Theta\right)$.

Proof. We have to verify conditions (1), (2) and (3) of Definition 1, § 9 for the prospective holomorphy type τ from E to $\mathscr{P}_\Theta({}^\ell E; F)$. Condition (1) is clearly satisfied.

As to Condition (2), if

$$P \in \mathscr{P}_\Theta({}^{\ell+0}E; F) \subset \mathscr{P}({}^\ell E; F),$$

then

$$\frac{1}{\ell!}\hat{d}^\ell P = P,$$

showing that

$$\frac{1}{\ell!}\hat{d}^\ell \mathscr{P}_\Theta({}^{\ell+0}E; F)$$

coincides with $\mathscr{P}_\Theta({}^\ell E; F)$ as a normed space.

To verify Condition (3), let $k \in \mathbb{N}$, $m \in \mathbb{N}$, $k \leqslant m$, $x \in E$ and

$$P \in \frac{1}{\ell!}\hat{d}^\ell \mathscr{P}_\Theta({}^{\ell+m}E; F)$$

be given. This means that

$$P = \frac{1}{\ell!}\hat{d}^\ell Q$$

for a unique $Q \in \mathscr{P}_\Theta({}^{\ell+m}E; F)$. We then have (Corollary 1, § 7)

$$\frac{1}{k!}\hat{d}^k P(x) = \frac{1}{k!}\hat{d}^k\left(\frac{1}{\ell!}\hat{d}^\ell Q\right)(x) = \frac{1}{\ell!}\hat{d}^\ell\left[\frac{1}{(\ell+k)!}\hat{d}^{\ell+k}Q(x)\right].$$

Since

$$\hat{d}^{\ell+k}Q(x) \in \mathscr{P}_\Theta({}^{\ell+k}E; F)$$

by Condition (3) of Definition 1, § 9, we see that

$$\hat{d}^k P(x) \in \frac{1}{\ell!}\hat{d}^\ell \mathscr{P}_\Theta({}^{\ell+k}E; F).$$

Moreover, we claim that

$$\left\| \frac{1}{k!} \hat{d}^k P(x) \right\|_\tau \leqslant (\sigma^{\ell+1})^m \cdot \|P\|_\tau \cdot \|x\|^{m-k},$$

where σ is associated with the holomorphy type Θ from E to F by Condition (3) of Definition 1, § 9. In fact, this inequality amounts to

$$\left\| \frac{1}{(\ell+k)!} \hat{d}^{\ell+k} Q(x) \right\|_\Theta \leqslant (\sigma^{\ell+1})^m \cdot \|Q\|_\Theta \cdot \|x\|^{m-k}.$$

This is true in case $m=0$, for then $k=0$, $Q \in \mathscr{P}_\Theta({}^\ell E; F)$ and

$$\frac{1}{\ell!} \hat{d}^\ell Q = Q.$$

This is also true in case $m \geqslant 1$. In fact, using Condition (3) of Definition 1, § 9 for Θ, we have

$$\left\| \frac{1}{(\ell+k)!} \hat{d}^{\ell+k} Q(x) \right\|_\Theta \leqslant \sigma^{\ell+m} \cdot \|Q\|_\Theta \cdot \|x\|^{m-k}.$$

It only remains to notice that $\sigma^{\ell+m} \leqslant \sigma^{\ell m+m} = (\sigma^{\ell+1})^m$. *Q. E. D.*

Proposition 2. *Let Θ be a holomorphy type from E to F, and*

$$\tau = \frac{1}{\ell!} \hat{d}^\ell \Theta$$

be the corresponding holomorphy type from E to $\mathscr{P}_\Theta({}^\ell E; F)$, where $\ell \in \mathbb{N}$ (see Proposition 1). *If $f \in \mathscr{H}_\Theta(U; F)$, then*

$$\hat{d}^\ell f \in \mathscr{H}_\tau(U; \mathscr{P}_\Theta({}^\ell E; F)).$$

Proof. Letting $\xi \in U$ and

$$P_m = \frac{1}{m!} \hat{d}^m f(\xi) \qquad (m \in \mathbb{N}),$$

choose some real number $\rho > 0$ such that $B_\rho(\xi) \subset U$, and such that

$$(1) \quad \frac{1}{\ell!}\hat{d}^\ell f(x) = \sum_{m=\ell}^{\infty} \frac{1}{\ell!}\hat{d}^\ell P_m(x-\xi) = \sum_{m=0}^{\infty} \frac{1}{\ell!}\hat{d}^\ell P_{\ell+m}(x-\xi)$$

for every $x \in B_\rho(\xi)$ and $\ell \in \mathbb{N}$, where the series is assumed to be convergent in the sense of $\mathscr{P}_\Theta({}^\ell E; F)$, by (3) of Lemma 1, § 9. Since we can use Conditions (1) and (2) of Definition 2, § 9, we see that

$$(2) \quad \begin{aligned} \left\| \frac{1}{\ell!}\hat{d}^\ell P_{\ell+m}(x-\xi) \right\|_\Theta &\leqslant \sigma^{\ell+m} \cdot \|P_{\ell+m}\|_\Theta \cdot \|x-\xi\|^m \\ &\leqslant C \cdot (\sigma c)^\ell \cdot (\sigma c)^m \cdot \|x-\xi\|^m, \end{aligned}$$

by virtue also of Condition (3) of Definition 1, § 9. It follows from (1) and (2) that

$$(3) \quad \hat{d}^\ell f \in \mathscr{H}(U; \mathscr{P}_\Theta({}^\ell E; F)).$$

Moreover we see that

$$(4) \quad \frac{1}{\ell!}\hat{d}^\ell P_{\ell+m} \in \frac{1}{\ell!}\hat{d}^\ell \mathscr{P}_\Theta({}^{\ell+m}E; F),$$

and that

$$(5) \quad \left\| \frac{1}{\ell!}\hat{d}^\ell P_{\ell+m} \right\|_\tau = \|P_{\ell+m}\|_\Theta \leqslant C \cdot c^\ell \cdot c^m.$$

What remains is to notice that (3), (4) and (5) complete the proof. *Q. E. D.*

Corollary 1. *Let $f \in \mathscr{H}_\Theta(U; F)$ and $B_\rho(\xi) \subset U$. Then*

$$\left\| \frac{1}{\ell!}\hat{d}^\ell f(x) \right\|_\Theta \leqslant \sum_{m=\ell}^{\infty} \sigma^m \rho^{m-\ell} \cdot \left\| \frac{1}{m!}\hat{d}^m f(\xi) \right\|_\Theta$$

for every $x \in B_\rho(\xi)$ and $\ell \in \mathbb{N}$.

Proof. By Proposition 2, we have

$$\frac{1}{\ell!}\hat{d}^{\ell} f \in \mathcal{H}(U; \mathcal{P}_{\Theta}(^{\ell}E; F)).$$

Hence the Taylor series of this function at ξ, namely

$$\frac{1}{\ell!}\hat{d}^{\ell} f(x) = \sum_{m=\ell}^{\infty} \frac{1}{\ell!}\hat{d}^{\ell} P_m(x-\xi),$$

where

$$P_m = \frac{1}{m!}\hat{d}^m f(\xi) \qquad (m \in \mathbb{N}),$$

converges to the indicated sum, in the sense of $\mathcal{P}_{\Theta}(^{\ell}E; F)$, for every $x \in B_{\rho}(\xi)$, by Proposition 1, § 7. By using Condition (3) of Definition 1, § 9, we get

$$\left\| \frac{1}{\ell!}\hat{d}^{\ell} f(x) \right\|_{\Theta} \leqslant \sum_{m=\ell}^{\infty} \sigma^m \|P_m\|_{\Theta}\, \rho^{m-\ell}$$

if $x \in B_{\rho}(\xi)$. *Q. E. D.*

Corollary 2. *Let $f \in \mathcal{H}_{\Theta}(U; F)$ and $B_{\rho}(X) \subset U$. Then*

$$\sum_{\ell=0}^{\infty} \varepsilon^{\ell} \cdot \sup_{x \in B_{\rho}(X)} \left\| \frac{1}{\ell!}\hat{d}^{\ell} f(x) \right\|_{\Theta} \leqslant \sum_{m=0}^{\infty} [\sigma(\rho+\varepsilon)]^m \cdot \sup_{x \in X} \left\| \frac{1}{m!}\hat{d}^m f(x) \right\|_{\Theta}$$

for every real number $\varepsilon > 0$.

Proof. Apply Corollary 1. *Q. E. D.*

§ 11. Topology on Spaces of Holomorphic Mappings

Lemma 1. *Let p be a seminorm on $\mathscr{H}_\Theta(U; F)$ and K be a compact subset of U. Then the following conditions are equivalent:*

(1) *Given any real number $\varepsilon > 0$, we can find a real number $c(\varepsilon) > 0$ such that*

$$p(f) \leqslant c(\varepsilon) \sum_{m=0}^{\infty} \varepsilon^m \cdot \sup_{x \in K} \left\| \frac{1}{m!} \hat{d}^m f(x) \right\|_\Theta$$

for every $f \in \mathscr{H}_\Theta(U; F)$.

(2) *Given any real number $\varepsilon > 0$ and any open subset V of U containing K, we can find a real number $c(\varepsilon, V) > 0$ such that*

$$p(f) \leqslant c(\varepsilon, V) \sum_{m=0}^{\infty} \varepsilon^m \cdot \sup_{x \in V} \left\| \frac{1}{m!} \hat{d}^m f(x) \right\|_\Theta$$

for every $f \in \mathscr{H}_\Theta(U; F)$.

Proof. It is clear that (1) implies (2). To prove that (2) implies (1), we apply Corollary 2 of § 10 by taking $X = K$ and by assuming further that $\rho \leqslant \varepsilon$. *Q. E. D.*

Definition 1. A seminorm p on $\mathscr{H}_\Theta(U; F)$ is said to be *ported* by a compact subset K of U if the equivalent conditions (1) and (2) of Lemma 1 hold. The natural topology $\mathscr{I}_{\omega,\Theta}$ on $\mathscr{H}_\Theta(U; F)$, is defined by the seminorms on $\mathscr{H}_\Theta(U; F)$ that are ported by compact subsets of U. It is plainly separated.

Remark 1. It is to be remarked that, by Proposition 2, § 9, once $f \in \mathscr{H}_\Theta(U; F)$ and a compact subset K of U are given, we can find a real number $\varepsilon > 0$ and an open subset V of U containing K such that

$$\sum_{m=0}^{\infty} \varepsilon^m \cdot \sup_{x \in V} \left\| \frac{1}{m!} \hat{d}^m f(x) \right\|_{\Theta} < \infty,$$

and, in particular,

$$\sum_{m=0}^{\infty} \varepsilon^m \cdot \sup_{x \in K} \left\| \frac{1}{m!} \hat{d}^m f(x) \right\|_{\Theta} < \infty.$$

These facts are to be compared, respectively, with conditions (2) and (1) in Lemma 1.

Proposition 1. *Let Λ be a set, $\mathscr{F}$ a filter on Λ, $(f_\lambda)_{\lambda \in \Lambda}$ a family of elements of $\mathscr{H}_\Theta(U;F)$ indexed by Λ and $f \in \mathscr{H}_\Theta(U;F)$. Assume that, corresponding to every compact subset K of U, we can find a real number $\varepsilon > 0$ such that*

$$\lim_{\lambda \to \mathscr{F}} \sum_{m=0}^{\infty} \varepsilon^m \cdot \sup_{x \in K} \left\| \frac{1}{m!} \hat{d}^m (f_\lambda - f)(x) \right\|_{\Theta} = 0.$$

Then

$$\lim_{\lambda \to \mathscr{F}} f_\lambda = f$$

in the sense of the topology $\mathscr{I}_{\omega,\Theta}$ on $\mathscr{H}_\Theta(U;F)$.

The proof is trivial.

Remark 2. From Corollary 2, § 10, it follows that the assumption in Proposition 1 is equivalent to requiring that, corresponding to every compact subset K of U, we can find a real number $\varepsilon > 0$ and an open subset V of U containing K such that

$$\lim_{\lambda \to \mathscr{F}} \sum_{m=0}^{\infty} \varepsilon^m \cdot \sup_{x \in V} \left\| \frac{1}{m!} \hat{d}^m (f_\lambda - f)(x) \right\|_{\Theta} = 0.$$

Proposition 2. *Let $f \in \mathscr{H}_\Theta(U;F)$, $\xi \in U$ and U be ξ-equilibrated. Then the Taylor series of f at ξ converges to f in the sense of the topology $\mathscr{I}_{\omega,\Theta}$ on $\mathscr{H}_\Theta(U;F)$.*

The proof will be based on the following lemma.

Lemma 2. *Let $f \in \mathscr{H}_\Theta(U;F)$; $\xi \in U$ and U be ξ-equilibrated. Then, given any compact subset K of U, there exist real numbers $\gamma (0 < \gamma < 1)$, $C \geqslant 0$ and $c \geqslant 0$, and an open subset V of U containing K, such that*

$$\sup_{x \in V} \left\| \frac{1}{m!} \hat{d}^m (f - \tau_{\ell,f,\xi})(x) \right\|_\Theta \leqslant C \cdot c^m \cdot \gamma^\ell$$

for every $\ell \in \mathbb{N}$, $m \in \mathbb{N}$.

Proof. By Proposition 2, § 9, choose an open subset W of U containing the given compact subset K of U, and real numbers $C \geqslant 0$ and $c \geqslant 0$, such that W is ξ-equilibrated and

$$\left\| \frac{1}{m!} \hat{d}^m f(x) \right\|_\Theta \leqslant C \cdot c^m$$

for every $x \in W$ and $m \in \mathbb{N}$. Next choose a real number $\rho > 1$ and an open subset V of W containing K such that $\lambda \in \mathbb{C}$, $|\lambda| \leqslant \rho$, $x \in V$ imply that $(1-\lambda)\xi + \lambda x \in W$. Use

$$\hat{d}^m f \in \mathscr{H}(U; \mathscr{P}_\Theta({}^m E; F)),$$

by Proposition 2, § 10, and apply Lemma 1, § 6, to $\hat{d}^m f$ to get

$$\|\hat{d}^m f(x) - \tau_{\ell,\hat{d}^m f,\xi}(x)\|_\Theta \leqslant \frac{C \cdot c^m \cdot m!}{\rho^\ell (\rho - 1)}$$

for $x \in V$, $\ell \in \mathbb{N}$ and $m \in \mathbb{N}$. Now (Corollary 2, § 7)

$$\tau_{\ell,\hat{d}^m f,\xi} = \hat{d}^m \tau_{\ell+m,f,\xi},$$

so that

$$\left\| \frac{1}{m!} \hat{d}^m (f - \tau_{\ell+m, f, \xi})(x) \right\|_{\Theta} \leq \frac{C \cdot c^m}{\rho^{\ell}(\rho - 1)}$$

for $x \in V$, $\ell \in \mathbb{N}$ and $m \in \mathbb{N}$. In other notation,

$$\left\| \frac{1}{m!} \hat{d}^m (f - \tau_{\ell, f, \xi})(x) \right\|_{\Theta} \leq \frac{C \cdot (\rho c)^m}{\rho^{\ell}(\rho - 1)} \tag{1}$$

for $x \in V$, $\ell \in \mathbb{N}$, $m \in \mathbb{N}$ and $\ell \geq m$. However, (1) remains true for $\ell < m$. In fact, then

$$\hat{d}^m \tau_{\ell, f, \xi} = 0;$$

and since $V \subset W$, we have

$$\left\| \frac{1}{m!} \hat{d}^m f(x) \right\|_{\Theta} \leq C \cdot c^m \leq \frac{C \cdot (\rho c)^m}{\rho^{\ell}(\rho - 1)}$$

for $x \in V$, because $\rho - 1 < \rho^{m-\ell}$. Thus (1) is true for $x \in V$, $\ell \in \mathbb{N}$ and $m \in \mathbb{N}$. The lemma is thus proved if we replace

$$\frac{C}{\rho - 1} \quad \text{and} \quad \rho c$$

by C and c respectively, and also set $\gamma = 1/\rho$. *Q. E. D.*

Proof of Proposition 2. Apply Lemma 2. Then corresponding to every compact subset K of U, we can find real numbers $\gamma (0 < \gamma < 1)$, $C \geq 0$ and $c \geq 0$, and an open subset V of U containing K, such that if the real number $\varepsilon > 0$ satisfies $\varepsilon c < 1$, then

$$\sum_{m=0}^{\infty} \varepsilon^m \cdot \sup_{x \in V} \left\| \frac{1}{m!} \hat{d}^m (f - \tau_{\ell, f, \xi})(x) \right\|_{\Theta} \leq \frac{C \gamma^{\ell}}{1 - \varepsilon c}$$

for $\ell \in \mathbb{N}$. Then apply Proposition 1 (compare with Remark 2).

Q. E. D.

Proposition 3. *Each inclusion mapping*

$$\mathscr{H}_\Theta(U;F) \subset \mathscr{H}(U;F)$$

is continuous for the corresponding topologies $\mathscr{I}_{\omega,\Theta}$ *and* $\mathscr{I}_\omega$.

Proof. Apply Proposition 1, § 9. *Q. E. D.*

Proposition 4. *Let* Θ *be a holomorphy type from* E *to* F *and*

$$\tau = \frac{1}{\ell!}\hat{d}^\ell \Theta$$

be the corresponding holomorphy type from E *to* $\mathscr{P}_\Theta({}^\ell E;F)$, *where* $\ell \in \mathbb{N}$ (see Proposition 1, § 10). *Then the linear mapping*

$$f \in \mathscr{H}_\Theta(U;F) \mapsto \frac{1}{\ell!}\hat{d}^\ell f \in \mathscr{H}_\tau(U;\mathscr{P}_\Theta({}^\ell E;F))$$

(see Proposition 2, § 10) *is continuous for the corresponding* $\mathscr{I}_{\omega,\Theta}$ *and* $\mathscr{I}_{\omega,\tau}$ *topologies.*

Proof. Let p be a seminorm on $\mathscr{H}_\tau(U;\mathscr{P}_\Theta({}^\ell E;F))$ ported by a compact subset K of U. Let the real number $c(\varepsilon)>0$ correspond to every real number $\varepsilon>0$ so that, for every $g \in \mathscr{H}_\tau(U;\mathscr{P}_\Theta({}^\ell E;F))$, we have

$$p(g) \leqslant c(\varepsilon) \sum_{m=0}^{\infty} \varepsilon^m \cdot \sup_{x\in K} \left\| \frac{1}{m!}\hat{d}^m g(x) \right\|_\tau .$$

If then $f \in \mathscr{H}_\Theta(U;F)$, we have (Corollary 1, § 7)

$$\left\| \frac{1}{m!}\hat{d}^m\left(\frac{1}{\ell!}\hat{d}^\ell f\right)(x)\right\|_\tau = \left\| \frac{1}{\ell!}\hat{d}^\ell\left[\frac{1}{(\ell+m)!}\hat{d}^{\ell+m} f(x)\right]\right\|_\tau$$

$$= \left\| \frac{1}{(\ell+m)!}\hat{d}^{\ell+m} f(x)\right\|_\Theta ,$$

and so

$$p\left(\frac{1}{\ell!}\hat{d}^{\ell} f\right) \leqslant c(\varepsilon) \sum_{m=0}^{\infty} \varepsilon^{m} \cdot \sup_{x \in K} \left\| \frac{1}{(\ell+m)!} \hat{d}^{\ell+m} f(x) \right\|_{\Theta}$$

$$\leqslant \frac{c(\varepsilon)}{\varepsilon^{\ell}} \sum_{m=0}^{\infty} \varepsilon^{m} \cdot \sup_{x \in K} \left\| \frac{1}{m!} \hat{d}^{m} f(x) \right\|_{\Theta}$$

proving the desired continuity. *Q. E. D.*

§ 12. Bounded Subsets

Proposition 1. *Each of the following equivalent conditions is necessary and sufficient for a subset $\mathscr{X}$ of $\mathscr{H}_\Theta(U;F)$ to be bounded for $\mathscr{I}_{\omega,\Theta}$:*

(1) *Corresponding to every $\xi \in U$, there are real numbers $C \geqslant 0$ and $c \geqslant 0$ such that*

$$\left\| \frac{1}{m!} \hat{d}^m f(\xi) \right\|_\Theta \leqslant C \cdot c^m$$

for every $m \in \mathbb{N}$ and $f \in \mathscr{X}$.

(2) *Corresponding to every compact subset K of U, there are real numbers $C \geqslant 0$ and $c \geqslant 0$ such that*

$$\left\| \frac{1}{m!} \hat{d}^m f(x) \right\|_\Theta \leqslant C \cdot c^m$$

for every $m \in \mathbb{N}$, $f \in \mathscr{X}$ and $x \in K$.

(3) *Corresponding to every compact subset K of U, there are real numbers $C \geqslant 0$ and $c \geqslant 0$, and an open subset V of U containing K, such that*

$$\left\| \frac{1}{m!} \hat{d}^m f(x) \right\|_\Theta \leqslant C \cdot c^m$$

for every $m \in \mathbb{N}$, $f \in \mathscr{X}$ and $x \in V$.

Proof. Let $\mathscr{X}$ be bounded for $\mathscr{I}_{\omega,\Theta}$. We shall then prove (2). If $K \subset U$ is compact and $\alpha_m \geqslant 0$ $(m \in \mathbb{N})$ are real numbers such that $(\alpha_m)^{1/m} \to 0$ as $m \to \infty$, we have correspondingly a seminorm p on $\mathscr{H}_\Theta(U; F)$ defined by

$$p(f) = \sum_{m=0}^{\infty} \alpha_m \cdot \sup_{x \in K} \left\| \frac{1}{m!} \hat{d}^m f(x) \right\|_\Theta$$

for $f \in \mathscr{H}_\Theta(U; F)$, by Proposition 2, § 9. It is immediate that p is ported by K, hence continuous for $\mathscr{I}_{\omega,\Theta}$. Therefore p is bounded on $\mathscr{X}$. Now it is classical that if $s_{m,\lambda} \geqslant 0$ are real numbers for $m \in \mathbb{N}$ and $\lambda \in \Lambda$, where Λ is a set, then

$$\sup_{\lambda \in \Lambda} \sum_{m=0}^{\infty} \alpha_m s_{m,\lambda} < \infty$$

holds true for every sequence $(\alpha_m)_{m \in \mathbb{N}}$ of positive real numbers such that $(\alpha_m)^{1/m} \to 0$ as $m \to \infty$ if and only if there are real numbers $C \geqslant 0$ and $c \geqslant 0$ such that

$$s_{m,\lambda} \leqslant C \cdot c^m$$

for every $m \in \mathbb{N}$ and $\lambda \in \Lambda$. Therefore, the fact that every seminorm p of the above form is bounded on $\mathscr{X}$ implies (2).

Conversely, it is clear that (2) implies that $\mathscr{X}$ is bounded for $\mathscr{I}_{\omega,\Theta}$.

The implications (3) $\Rightarrow$ (2) $\Rightarrow$ (1) are clear too.

Let us finally prove (1) $\Rightarrow$ (3). In fact, if $\xi \in U$ and $C \geqslant 0$ and $c \geqslant 0$ correspond by (1), we may apply Corollary 1, § 10 to get

$$\left\| \frac{1}{\ell!} \hat{d}^\ell f(x) \right\|_\Theta \leqslant \frac{C}{1 - \sigma \rho c} (\sigma c)^\ell$$

for every $x \in B_\rho(\xi)$, $\ell \in \mathbb{N}$ and $f \in \mathscr{X}$, provided we choose $\rho > 0$ so that $B_\rho(\xi) \subset U$ and $\sigma \rho c < 1$. This suffices to show that (3) holds.

Q. E. D.

Definition 1. Corresponding to every compact subset K of U and every $m \in \mathbb{N}$, we have the seminorm p on $\mathscr{H}_\Theta(U; F)$ defined by

$$p(f) = \sup_{x \in K} \|\hat{d}^m f(x)\|_\Theta$$

for $f \in \mathscr{H}_\Theta(U;F)$. The $\mathscr{I}_\infty$ topology on $\mathscr{H}_\Theta(U;F)$ is defined by all such seminorms. Clearly $\mathscr{I}_\infty \subset \mathscr{I}_{\omega,\Theta}$ and $\mathscr{I}_\infty$ is separated.

Proposition 2. *On every $\mathscr{I}_{\omega,\Theta}$-bounded subset $\mathscr{X}$ of $\mathscr{H}_\Theta(U;F)$, the uniform structures associated with $\mathscr{I}_{\omega,\Theta}$ and $\mathscr{I}_\infty$ induce the same uniform structure. In particular, $\mathscr{I}_{\omega,\Theta}$ and $\mathscr{I}_\infty$ induce on $\mathscr{X}$ the same topology.*

Proof. Let us assume first that $0 \in \mathscr{X}$ and prove that a subset of $\mathscr{X}$ is a neighborhood of 0 in the topology on $\mathscr{X}$ induced by $\mathscr{I}_{\omega,\Theta}$ if and only if it is a neighborhood of 0 in the topology on $\mathscr{X}$ induced by $\mathscr{I}_\infty$. One half of this assertion is clear from $\mathscr{I}_\infty \subset \mathscr{I}_{\omega,\Theta}$. Conversely, let p be a $\mathscr{I}_{\omega,\Theta}$-continuous seminorm on $\mathscr{H}_\Theta(U;F)$. Assume that p is ported by a compact subset K of U, and let $c(\varepsilon)$ be as described in (1) of Lemma 1, § 11. Since $\mathscr{X}$ is $\mathscr{I}_{\omega,\Theta}$-bounded, there are C and c according to Condition (2) of Proposition 1. Next choose $\varepsilon > 0$ so that $\varepsilon c < 1$ and $\mu \in \mathbb{N}$ by

$$C \cdot c(\varepsilon) \sum_{m > \mu} (\varepsilon c)^m \leqslant \tfrac{1}{2}.$$

Define the $\mathscr{I}_\infty$-continuous seminorm q by

$$q(f) = c(\varepsilon) \sum_{m=0}^{\mu} \varepsilon^m \cdot \sup_{x \in K} \left\| \frac{1}{m!} \hat{d}^m f(x) \right\|_\Theta .$$

It is then clear that, if $f \in \mathscr{X}$ and $q(f) \leqslant \frac{1}{2}$, then $p(f) \leqslant 1$. This proves the remaining half of the above assertion.

If we next consider any subset $\mathscr{X}$ bounded for $\mathscr{I}_{\omega,\Theta}$, the set $\mathscr{X} - \mathscr{X}$ of all differences of two elements of $\mathscr{X}$ is bounded for $\mathscr{I}_{\omega,\Theta}$ and contains 0. Since the neighborhoods of 0 in the topologies on $\mathscr{X} - \mathscr{X}$ induced by $\mathscr{I}_{\omega,\Theta}$ and $\mathscr{I}_\infty$ are identical, it follows that the uniform structures on $\mathscr{X}$ induced by the uniform structures associated to $\mathscr{I}_{\omega,\Theta}$ and $\mathscr{I}_\infty$ are identical too. *Q.E.D.*

Corollary 1. *If* $f_\ell \in \mathscr{H}_\Theta(U;F)$ *for* $\ell \in \mathbb{N}$, *and* $f \in \mathscr{H}_\Theta(U;F)$, *then* $f_\ell \to f$ *for* $\mathscr{I}_{\omega,\Theta}$ *as* $\ell \to \infty$ *if and only if* $(f_\ell)_{\ell \in \mathbb{N}}$ *is bounded for* $\mathscr{I}_{\omega,\Theta}$ *and* $f_\ell \to f$ *for* $\mathscr{I}_\infty$ *as* $\ell \to \infty$.

Proof. If $f_\ell \to f$ for $\mathscr{I}_{\omega,\Theta}$ as $\ell \to \infty$, then clearly $(f_\ell)_{\ell \in \mathbb{N}}$ is $\mathscr{I}_{\omega,\Theta}$-bounded and also $f_\ell \to f$ for $\mathscr{I}_\infty$ as $\ell \to \infty$.

Conversely, let us assume that $(f_\ell)_{\ell \in \mathbb{N}}$ is $\mathscr{I}_{\omega,\Theta}$-bounded and that $f_\ell \to f$ for $\mathscr{I}_\infty$ as $\ell \to \infty$. To prove that $f_\ell \to f$ for $\mathscr{I}_{\omega,\Theta}$ as $\ell \to \infty$ it suffices to notice that the subset of $\mathscr{H}_\Theta(U;F)$ formed by all $f_\ell (\ell \in \mathbb{N})$, and by f, is $\mathscr{I}_{\omega,\Theta}$-bounded, and then to apply Proposition 2. *Q. E. D.*

Proposition 3. *Each subset* $\mathscr{X}$ *of* $\mathscr{H}_\Theta(U;F)$ *bounded for* $\mathscr{I}_{\omega,\Theta}$ *is equicontinuous at every point of* U.

Proof. Let $\xi \in U$ and C and c correspond to it by Condition (1) of Proposition 1. We have

$$f(x) = \sum_{m=0}^{\infty} \frac{1}{m!} \hat{d}^m f(\xi)(x-\xi)$$

for every $f \in \mathscr{H}(U;F)$ provided $x \in B_\rho(\xi) \subset U$. Using Proposition 1, § 9, we get

$$\begin{aligned} \|f(x)-f(\xi)\| &\leqslant \sum_{m=1}^{\infty} \left\| \frac{1}{m!} \hat{d}^m f(\xi) \right\| \cdot \|x-\xi\|^m \\ &\leqslant \sum_{m=1}^{\infty} \sigma^m \cdot \left\| \frac{1}{m!} \hat{d}^m f(\xi) \right\|_\Theta \cdot \|x-\xi\|^m \\ &\leqslant \sum_{m=1}^{\infty} \sigma^m \cdot C \cdot c^m \cdot \|x-\xi\|^m = \frac{C\sigma c\|x-\xi\|}{1-\sigma c\|x-\xi\|} \end{aligned}$$

provided $x \in B_\rho(\xi)$, $\sigma c \rho \leqslant 1$ and $f \in \mathscr{X}$, from which equicontinuity follows. *Q. E. D.*

Remark 1. Proposition 3 for arbitrary Θ follows simply from the particular case in which Θ is the current holomorphy type (see § 14), in view of Proposition 3, § 11.

§ 13. Relatively Compact Subsets

Definition 1. A subset $\mathscr{X}$ of $\mathscr{H}_\Theta(U;F)$ is said to be *relatively compact at a point* $\xi \in U$ if, for every $m \in \mathbb{N}$, the set

$$\{\hat{d}^m f(\xi) | f \in \mathscr{X}\}$$

is relatively compact in $\mathscr{P}_\Theta(^m E;F)$.

Proposition 1. *A subset $\mathscr{X}$ of $\mathscr{H}_\Theta(U;F)$ is relatively compact for $\mathscr{I}_{\omega,\Theta}$ if and only if $\mathscr{X}$ is bounded for $\mathscr{I}_{\omega,\Theta}$, and $\mathscr{X}$ is relatively compact at every point of U.*

Proof. Assume that $\mathscr{X}$ is relatively compact for $\mathscr{I}_{\omega,\Theta}$. Then $\mathscr{X}$ is bounded for $\mathscr{I}_{\omega,\Theta}$. Moreover the mapping

$$f \in \mathscr{H}_\Theta(U;F) \mapsto \hat{d}^m f(\xi) \in \mathscr{P}_\Theta(^m E;F)$$

is continuous, for every $\xi \in U$ and $m \in \mathbb{N}$. Hence the image of $\mathscr{X}$ is relatively compact in $\mathscr{P}_\Theta(^m E;F)$ for every $m \in \mathbb{N}$, that is, $\mathscr{X}$ is relatively compact at every $\xi \in U$.

Conversely, let us assume that $\mathscr{X}$ is bounded for $\mathscr{I}_{\omega,\Theta}$ and that $\mathscr{X}$ is relatively compact at every point of U. By Proposition 2, § 12, the closure of $\mathscr{X}$ in the topology $\mathscr{I}_{\omega,\Theta}$ coincides with the closure of $\mathscr{X}$ in the topology $\mathscr{I}_\infty$. Moreover since such a closure is $\mathscr{I}_{\omega,\Theta}$-bounded, the topologies induced on it by $\mathscr{I}_{\omega,\Theta}$ and by $\mathscr{I}_\infty$ are identical. Hence, in order to prove that $\mathscr{X}$ is relatively compact for $\mathscr{I}_{\omega,\Theta}$, we shall show that $\mathscr{X}$ is relatively compact for $\mathscr{I}_\infty$.

To this end, consider the cartesian product

$$S = \prod_{m=0}^{\infty} \mathscr{P}_{\Theta}(^m E; F) = F_{\Theta}[[E]]$$

(vector space of all formal power series from E to F) endowed with its cartesian product topology. We define a natural mapping Φ from $\mathscr{H}_{\Theta}(U;F)$ into the vector space $\mathscr{C}(U;S)$ of all continuous S-valued functions on U by associating with every $f \in \mathscr{H}_{\Theta}(U;F)$ the function $\Phi(f)$ defined at every point $x \in U$ by

$$\Phi(f)(x) = \left(\frac{1}{m!}\hat{d}^m f(x)\right)_{m \in \mathbb{N}} \in S.$$

It is immediate that $\Phi(f) \in \mathscr{C}(U;S)$, because each $\hat{d}^m f$ is holomorphic, hence continuous, from U to $\mathscr{P}_{\Theta}(^m E; F)$. The mapping Φ is linear and one-to-one. It is a homeomorphism if we endow $\mathscr{H}_{\Theta}(U;F)$ with the $\mathscr{I}_{\infty}$ topology and $\mathscr{C}(U;S)$ with the compact-open topology. By the assumption that $\mathscr{X}$ is relatively compact at every point x of U, we see that the set

$$\{\Phi(f)(x) \mid f \in \mathscr{X}\}$$

is relatively compact in S. Therefore, in view of Ascoli's theorem, in order to show that the image set $\Phi(\mathscr{X})$ is relatively compact in $\mathscr{C}(U;S)$, we must show that it is equicontinuous at every point of U. In other words, we must show that, for every $m \in \mathbb{N}$, the subset

$$\{\hat{d}^m f \mid f \in \mathscr{X}\}$$

of $\mathscr{C}(U;\mathscr{P}_{\Theta}(^m E;F))$ is equicontinuous. This follows from the fact that $\mathscr{X}$ is bounded for $\mathscr{I}_{\omega,\Theta}$, from Proposition 4, § 11 and from Proposition 3, § 12. *Q. E. D.*

Proposition 2. *If U is connected, a subset $\mathscr{X}$ of $\mathscr{H}_{\Theta}(U;F)$ is relatively compact for $\mathscr{I}_{\omega,\Theta}$ if and only if $\mathscr{X}$ is bounded for $\mathscr{I}_{\omega,\Theta}$, and $\mathscr{X}$ is relatively compact at a single point of U.*

The proof will be based on the following lemma.

Lemma 1. *Let* $\mathscr{X}$ *be a subset of* $\mathscr{H}_\Theta(U;F)$ *such that*

(1) $\mathscr{X}$ *is relatively compact at some* $\xi \in U$, *and*

(2) *There exist real numbers* $\rho > 0$ *and* $C \geqslant 0$ *such that* $B_\rho(\xi) \subset U$ *and*

$$\left\| \frac{1}{m!} \hat{d}^m f(\xi) \right\|_\Theta \leqslant \frac{C}{(\sigma \rho)^m}$$

for every $f \in \mathscr{X}$ *and* $m \in \mathbb{N}$.

Then $\mathscr{X}$ *is relatively compact at every point of* $B_\rho(\xi)$.

Proof. For every $f \in \mathscr{H}_\Theta(U;F)$ set

$$P_{m,f} = \frac{1}{m!} \hat{d}^m f(\xi) \qquad (m \in \mathbb{N}).$$

We then have

$$\hat{d}^\ell f(x) = \sum_{m=\ell}^{\infty} \hat{d}^\ell P_{m,f}(x-\xi)$$

for $x \in B_\rho(\xi)$ and $\ell \in \mathbb{N}$, convergence being in the sense of $\mathscr{P}_\Theta({}^\ell E;F)$, by Proposition 2, § 10. Using (2) and Condition (3) of Definition 1, § 9, we have

$$\|\hat{d}^\ell P_{m,f}(x-\xi)\|_\Theta \leqslant \sigma^m \cdot \frac{C}{(\sigma \rho)^m} \cdot \|x-\xi\|^{m-\ell}$$

for every $f \in \mathscr{X}$, $\ell \in \mathbb{N}$, $m \in \mathbb{N}$, $\ell \leqslant m$. Using this estimate and the above series, we see that $\mathscr{X}$ is relatively compact at every point x such that $\|x-\xi\| < \rho$. *Q.E.D.*

Proof of Proposition 2. Assume that $\mathscr{X}$ is bounded for $\mathscr{I}_{\omega,\Theta}$. Let X be the set of points of U where $\mathscr{X}$ is relatively compact. Lemma 1 shows immediately that X is open. We now show that

X is closed. Let $\eta \in U$ be in the closure of X. Since $\mathscr{X}$ is $\mathscr{I}_{\omega,\Theta}$-bounded, there are $\rho > 0$ and $C \geqslant 0$ such that $B_\rho(\eta) \subset U$ and

$$\left\| \frac{1}{m!} \hat{d}^m f(x) \right\|_\Theta \leqslant \frac{C}{(\sigma \rho)^m}$$

for every $f \in \mathscr{X}$, $m \in \mathbb{N}$ and $x \in B_\rho(\eta)$, by Proposition 1, Condition (3), § 12. Let $r = \rho/2$, $\xi \in X \cap B_r(\eta)$. Then the fact that $\xi \in X$ and $\xi \in B_\rho(\eta)$, hence that

$$\left\| \frac{1}{m!} \hat{d}^m f(\xi) \right\|_\Theta \leqslant \frac{C}{(\sigma \rho)^m} \leqslant \frac{C}{(\sigma r)^m}$$

for every $f \in \mathscr{X}$ and $m \in \mathbb{N}$, implies that $\mathscr{X}$ is relatively compact at every point of $B_r(\xi)$, by Lemma 1, since $B_r(\xi) \subset B_\rho(\eta) \subset U$. Now $\eta \in B_r(\xi)$. Hence $\eta \in X$, proving that X is closed in U. The lemma then follows by connectedness of U. *Q.E.D.*

We shall not discuss further results concerning relative compactness, equicontinuity, etc., which follows from the results already proved and known general principles. Let us, however, make the following remark.

Remark 1. Proposition 2 of § 12 can be improved as follows. Let X be a subset of U. Denote by $\mathscr{I}_{\infty,X}$ the topology on $\mathscr{H}_\Theta(U;F)$ defined by the family formed by each of the following seminorms

$$p(f) = \|\hat{d}^m f(x)\|_\Theta$$

for $f \in \mathscr{H}_\Theta(U;F)$, where $x \in X$ and $m \in \mathbb{N}$. Then it follows from Proposition 2 and 3 of § 12, and from general principles concerning equicontinuity, that if X is dense in U, then $\mathscr{I}_{\omega,\Theta}$ and $\mathscr{I}_{\infty,X}$ induce the same topology on every $\mathscr{I}_{\omega,\Theta}$-bounded subset; and analogously for uniform structures. However the above proof of Propositions 1 and 2 will show that the same conclusion about the identity of $\mathscr{I}_{\omega,\Theta}$ and $\mathscr{I}_{\infty,X}$ on every $\mathscr{I}_{\omega,\Theta}$-bounded subset, as well as of their uniform structures, remains true if X intersects every connected component of U. This in turn could be used to provide short proofs of the above Propositions 1 and 2. A corresponding remark applies to Corollary 1, § 12.

§ 14. The Current Holomorphy Type

Definition 1. The *current* holomorphy type from E to F is the holomorphy type Θ for which $\mathscr{P}_\Theta({}^m E; F) = \mathscr{P}({}^m E; F)$, as normed vector spaces, for every $m \in \mathbb{N}$. Then $\mathscr{H}_\Theta(U; F) = \mathscr{H}(U; F)$, and $\mathscr{I}_{\omega,\Theta}$ is denoted by $\mathscr{I}_\omega$.

Certain of the preceding considerations simplify for the current holomorphy type, as we shall try to explain in the present section. It will be tacitly assumed that all considerations on $\mathscr{H}(U; F)$ are taken with respect to the current holomorphy type.

Lemma 1. *A seminorm p on $\mathscr{H}(U; F)$ is ported by a compact subset K of U if and only if, given any open subset V of U containing K, we can find a real number $c(V) > 0$ such that*

$$p(f) \leqslant c(V) \cdot \sup_{x \in V} \| f(x) \|$$

for every $f \in \mathscr{H}(U; F)$.

Proof. Let p be ported by K in the sense of Definition 1 of § 11, so that there exists a real number $c(\varepsilon) > 0$ corresponding to every real number $\varepsilon > 0$, such that

$$p(f) \leqslant c(\varepsilon) \sum_{m=0}^{\infty} \varepsilon^m \sup_{x \in K} \left\| \frac{1}{m!} \hat{d}^m f(x) \right\|$$

holds for every $f \in \mathscr{H}(U; F)$. Now, given any open subset V of U containing K, choose a real number $\rho > 0$ in such a way that any closed ball of radius ρ whose center lies in K will be contained in V. By Cauchy's inequality, we shall have

$$\sup_{x \in K} \left\| \frac{1}{m!} \hat{d}^m f(x) \right\| \leqslant \frac{1}{\rho^m} \sup_{x \in V} \| f(x) \| .$$

Take $\varepsilon = \rho/2$ and set $c(V) = 2c(\varepsilon)$ to conclude that p is ported by K in the sense of the statement of Lemma 1.

The converse is obvious, by Condition (2) of Lemma 1, § 11.

Q. E. D.

Remark 1. The topology $\mathcal{I}_\omega$ on $\mathcal{H}(U; F)$ is therefore defined by the seminorms on it that are ported by compact subsets of U in the sense of the statement of Lemma 1.

In the case of boundedness in the sense of $\mathcal{I}_\omega$ for subsets of $\mathcal{H}(U; F)$, besides the three equivalent conditions provided by Proposition 1, § 12, we have also the following ones.

Proposition 1. *Each of the following equivalent conditions is necessary and sufficient for a subset $\mathcal{X}$ of $\mathcal{H}(U; F)$ to be bounded for $\mathcal{I}_\omega$:*

(1) *Corresponding to every compact subset K of U, there is a real number $C \geqslant 0$ such that*

$$\| f(x) \| \leqslant C$$

for every $f \in \mathcal{X}$ and $x \in K$.

(2) *Corresponding to every compact subset K of U, there are a real number $C \geqslant 0$ and an open subset V of U containing K such that*

$$\| f(x) \| \leqslant C$$

for every $f \in \mathcal{X}$ and $x \in V$.

Proof. If (2) holds true, then $\mathscr{X}$ is bounded for $\mathscr{I}_\omega$, by Remark 1. If $\mathscr{X}$ is bounded for $\mathscr{I}_\omega$, then (1) is true. As a matter of fact, the seminorm p given on $\mathscr{H}(U;F)$ by

$$p(f) = \sup_{x \in K} \| f(x) \|$$

for all $f \in \mathscr{H}(U;F)$ is obviously continuous for $\mathscr{I}_\omega$; hence p must be bounded on $\mathscr{X}$. Finally (1) implies (2). In fact, (1) means that $\mathscr{X}$ is bounded for the topology induced on $\mathscr{H}(U;F)$ by the compact-open topology on the vector space $\mathscr{C}(U;F)$ of all continuous F-valued functions on U. Since U is metrizable, bounded subsets of $\mathscr{C}(U;F)$ have the property expressed by (2). *Q. E. D.*

§ 15. Bibliographical References

The basic facts of the theory of holomorphic mappings between Banach spaces, developed in § 4 through § 7, are to be found in [12] and [2], for instance.

We refer to [13] for the concept of the topology $\mathcal{I}_\omega$ on $\mathcal{H}(U; F)$ and its properties treated in § 8 and § 14. The notion of a seminorm on $\mathcal{H}(U; F)$ ported by a compact subset of U extends, and is analogous to, the notion of a continuous linear form (analytic functional) on $\mathcal{H}(U; \mathbb{C})$ ported by a compact subset of U, as defined in [10] when E is finite dimensional. Under such a restriction on E, one usually starts by introducing the natural topology on $\mathcal{H}(U; \mathbb{C})$ as the one induced on it by the compact-open topology on $\mathcal{C}(U; \mathbb{C})$. Next one shows that the linear forms on $\mathcal{H}(U; \mathbb{C})$ that are continuous are those ported by compact subsets of U. Conversely, the idea of [13] is to start from the more general notion of a seminorm on $\mathcal{H}(U; F)$ ported by a compact subset of U and use all such seminorms to define the natural topology $\mathcal{I}_\omega$ on $\mathcal{H}(U; F)$ without restricting E to be finite dimensional.

[1, 7] and [4] are the classical references on the topology of spaces of holomorphic functions; [11] is a more recent contribution to the subject, but still limited to the finite dimensional case. Classically, one assumes E to be finite dimensional and at first introduces the natural topology on $\mathcal{H}(U; \mathbb{C})$ as the one induced on it by the compact-open topology on $\mathcal{C}(U; \mathbb{C})$, where U is open in E. Next, K being a compact subset of E, one defines the space $\mathcal{H}(K; \mathbb{C})$ of all germs of holomorphic functions on K, the natural topology on $\mathcal{H}(K; \mathbb{C})$ being obtained as the inductive

limit of the natural topology on $\mathscr{H}(U; \mathbb{C})$, for all open subsets U of E containing the compact subset K of E. By dropping the assumption that E is finite dimensional, we comment without further explanation that the definition of $\mathscr{I}_\omega$ on $\mathscr{H}(U; F)$ corresponds to, conversely, introducing at first the natural topology on $\mathscr{H}(K; F)$ for K a compact subset of E, and next looking at the natural topology on $\mathscr{H}(U; F)$, where U is an open subset of E, as the projective limit of the natural topology on $\mathscr{H}(K; F)$, for all compact subsets K of U.

The concept of a holomorphy type Θ from E to F and that of the topology $\mathscr{I}_{\omega,\Theta}$ on $\mathscr{H}_\Theta(U; F)$, as developed in § 9 through § 13, was treated in [14]. Its motivation lies in previous work on nuclearly entire functions [6, 15], and in the various known types of continuous multilinear mappings [5], as well as their analogues for continuous homogeneous polynomials.

For the definition and applications of Banach manifolds, we refer the reader to [8, 16, 3] and [9].

We next list the bibliography quoted in this monograph.

1. Dias, C. L. S.: Espaços vetoriais topológicos e sua aplicação nos espaços funcionais analíticos. Boletim da Sociedade de Matemática de São Paulo **5,** 1–58 (1952).
2. Douady, A.: Le problème des modules pour les sous-espaces analytiques compacts d'un espace analytique donné. Annales de l'Institut Fourier **XVI,** 1–95 (1966).
3. Eells, J.: A setting for global analysis. Bulletin of the American Mathematical Society **72,** 751–807 (1966).
4. Grothendieck, A.: Sur certains espaces de fonctions holomorphes I, II, Journal für die Reine und Angewandte Mathematik **192,** 35–64, 77–95 (1953).
5. —, Produits tensoriels topologiques et espaces nucléaires. Memoirs of the American Mathematical Society **16,** 1–140 (1955).
6. Gupta, C. P.: Malgrange theorem for nuclearly entire functions of bounded type on a Banach space. Instituto de Matemática Pura

e Aplicada, Rio de Janeiro, Notas de Matemática **37,** 1–50 (1968).

7. Köthe, G.: Dualität in der Funktionentheorie. Journal für die Reine und Angewandte Mathematik **191,** 30–49 (1953).
8. Lang, S.: Introduction to differentiable manifolds. Interscience Publishers, USA (1962).
9. Lazard, M.: Mimeographed notes of a book on Banach manifolds and Lie groups (to appear).
10. Martineau, A.: Sur les fonctionnelles analytiques et la transformation de Fourier-Borel. Journal d'Analyse Mathématique **XI,** 1–164 (1963).
11. —, Sur la topologie des espaces de fonctions holomorphes. Mathematische Annalen **163,** 62–88 (1966).
12. Nachbin, L.: Lectures on the theory of distributions. University of Rochester (1963). Reproduced by Universidade do Recife. Textos de Matemática **15,** 1–280 (1964).
13. —, On the topology of the space of all holomorphic functions on a given open subset. Indagationes Mathematicae, Koninklijke Nederlandse Akademie van Wetenschappen, Proceedings, Series A **LXX,** 366–368 (1967).
14. —, On spaces of holomorphic functions of a given type. Proceedings of the Conference on Functional Analysis, University of California at Irvine, USA 50–60 (1966). Thompson Book Company (1967).
15. —, and C. P. Gupta: On Malgrange theorem for nuclearly entire functions (to appear).
16. Palais, R. S.: Lectures on the differential topology of infinite dimensional manifolds. Brandeis University, 1–125 (1965).